Sources and Sinks of Filtered Total Mercury and Concentrations of Total Mercury of Solids and of Filtered Methylmercury, Sinclair Inlet, Kitsap County, Washington, 2007–10

By Anthony J. Paulson, Richard S. Dinicola, Marlene A. Noble, Richard J. Wagner, Raegan L. Huffman, Patrick W. Moran, and John F. DeWild

Prepared in cooperation with Department of the Navy
Naval Facilities Engineering Command, Northwest

Scientific Investigations Report 2012-5223

U.S. Department of the Interior
U.S. Geological Survey

U.S. Department of the Interior
KEN SALAZAR, Secretary

U.S. Geological Survey
Marcia K. McNutt, Director

U.S. Geological Survey, Reston, Virginia: 2012

For more information on the USGS—the Federal source for science about the Earth, its natural and living resources, natural hazards, and the environment, visit http://www.usgs.gov or call 1–888–ASK–USGS.

For an overview of USGS information products, including maps, imagery, and publications, visit http://www.usgs.gov/pubprod

To order this and other USGS information products, visit http://store.usgs.gov

Suggested citation:
Paulson, A.J., Dinicola, R.S., Noble, M.A., Wagner, R.J., Huffman, R.L., Moran, P.W., and DeWild, J.F., 2012, Sources and sinks of filtered total mercury and concentrations of total mercury of solids and of filtered methylmercury, Sinclair Inlet, Kitsap County, Washington, 2007–10: U.S. Geological Survey Scientific Investigations Report 2012-5223, 94 p.

Contents

Contents—Continued

Contents—Continued

Figures

Figures—Continued

Tables

Conversion Factors, Datums, and Abbreviations and Acronyms

Conversion Factors

Inch/Pound to SI

Multiply	By	To obtain
Length		
foot (ft)	0.3048	meter (m)
inch (in.)	2.54	centimeter (cm)

SI to Inch/Pound

Multiply	By	To obtain
Length		
micrometer (μm)	0.003937	inch (in.)
millimeter (mm)	0.03937	inch (in.)
centimeter (cm)	0.3937	inch (in.)
meter (m)	3.281	foot (ft)
kilometer (km)	0.6214	mile (mi)
kilometer (km)	0.5400	mile, nautical (nmi)
meter (m)	1.094	yard (yd)
Area		
square kilometer (km^2)	247.1	acre
square meter (m^2)	10.76	square foot (ft^2)
square kilometer (km^2)	0.3861	square mile (mi^2)
Volume		
milliliter (mL)	0.0338	ounce, fluid (fl. oz)
cubic meter (m^3)	0.0002642	million gallons (Mgal)
cubic meter (m^3)	0.0008107	acre-foot (acre-ft)
liter (L)	1.057	quart (qt)
liter (L)	0.2642	gallon (gal)
Flow rate		
milliliter per minute (mL/min)	5.8857	cubic foot per second (ft^3/s)
cubic meter per second (m^3/s)	70.07	acre-foot per day (acre-ft/d)
liter per second (L/s)	15.85	gallon per minute (gal/min)
Mass		
gram	0.03527	ounce, avoirdupois (oz)
kilogram (kg)	2.205	pound avoirdupois (lb)
metric ton per year	1.102	ton per year (ton/yr)
Sedimentation rate		
gram per square centimeter (g cm^{-2})	0.228	ounce per square inch (oz/in^2)

Conversion Factors, Datums, and Abbreviations and Acronyms—Continued

Conversion Factors—Continued

Temperature in degrees Celsius (°C) may be converted to degrees Fahrenheit (°F) as follows:

$$°F=(1.8×°C)+32.$$

Temperature in degrees Fahrenheit (°F) may be converted to degrees Celsius (°C) as follows:

$$°C=(°F-32)/1.8.$$

Specific conductance is given in microsiemens per centimeter at 25 degrees Celsius (µS/cm at 25 °C).

Concentrations of chemical constituents in water are given either in milligrams per liter (mg/L), micrograms per liter (µg/L), or nanograms per liter (ng/L).

Concentrations of chemical constituents of solids are given in either percentage of dry weight, milligrams per kilogram (mg/kg) or nanograms per milligram (ng/mg), which are equivalent.

Datums

Vertical coordinate information is referenced to the North American Vertical Datum of 29 (NAVD 29). The vertical datum mean lower low water (MLLW) is defined as -1.83 m relative to NAVD 29.

Horizontal coordinate information is referenced to North American Datum of 1988 (NAD 88).

Altitude, as used in this report, refers to distance above the vertical datum.

Abbreviations and Acronyms

ADCP	acoustic Doppler current profiler
BNC	Bremerton naval complex
CERCLA	Comprehensive Environmental Response, Compensation, and Liability Act
CH3D	Curvilinear Hydrodynamics in 3-Dimensions
CSO	Combined sewer overflow
CTD	conductivity temperature depth
DOC	dissolved organic carbon
ENVVEST	ENVironmental inVESTment
F_{FW}	volume fraction of freshwater
F_{SW}	volume fraction of seawater
FC	fecal coliform
FMHg	filtered methylmercury
FTHg	filtered total mercury
HDPE	high-density polyethylene
HSPF	Hydrologic Simulation Program - FORTRAN
IAS	Initial Assessment Study
LMK	Label used by Kitsap County, Washington
LTMP	Long-term Monitoring Program
MHg	methylmercury
MLLW	mean lower low water
MW	monitoring well

Conversion Factors, Datums, and Abbreviations and Acronyms—Continued

Abbreviations and Acronyms—Continued

NBK Bremerton	Naval Base Kitsap at Bremerton
NPDES	National Pollutant Discharge Elimination System
NWQL	National Water Quality Laboratory
OU	Operable unit
OU NSC	Operable Unit Naval Supply Center
OUBT	Operable Unit B Terrestrial
PAHs	polyaromatic hydrocarbons
^{210}Pb	isotope of lead having a atomic weight of 210
PBDEs	Polybrominated diphenyl esters
PCBs	polychlorinated biphenyls
PDT	Pacific Daylight Savings Time
PETG	polyethylene terephthalate copolyester
PFA	Perfluoroalkoxy
PN	particulate nitrogen
PO	Port Orchard
PO-BLVD	Port Orchard Boulevard
POC	particulate organic carbon
PS02	Site 2 of the Preliminary Assessment
PSNS	Puget Sound Naval Shipyard
PSNS&IMF	Puget Sound Naval Shipyard and Intermediate Maintenance Facility
PSS	Practical salinity scale (unitless)
PTFE	Polytetrafluoroethylene
PTHg	particulate total mercury
PVC	polyvinyl chloride
QFF	quartz fiber filter
Q_{FW}	discharge of freshwater into Sinclair Inlet
Q_{SW}	transport of central Puget Sound seawater into Sinclair Inlet
RI/FS	Revisit Remedial Investigation/Feasibility Study
ROD	Record of Decision
RPD	Relative percent difference
S_{SI}	salinity of Sinclair Inlet
S_{DD}	salinity of dry dock discharge
S_{LL}	salinity of lower layer
S_{UL}	salinity of upper layer
SI	Sinclair Inlet
SPAWAR	The Space and Naval Warfare Systems Command
STHg	total mercury concentration of sediment
THg	total mercury
TSS	total suspended solids
WAWSC	Washington Water Science Center
WMRL	Wisconsin Mercury Research Laboratory
WTHg	total mercury in whole (unfiltered) water

Sources and Sinks of Filtered Total Mercury and Concentrations of Total Mercury of Solids and of Filtered Methylmercury, Sinclair Inlet, Kitsap County, Washington, 2007–10

By Anthony J. Paulson, Richard S. Dinicola, Marlene A. Noble, Richard J. Wagner, Raegan L. Huffman, Patrick W. Moran, and John F. DeWild

Abstract

The majority of filtered total mercury in the marine water of Sinclair Inlet originates from salt water flowing from Puget Sound. About 420 grams of filtered total mercury are added to Sinclair Inlet each year from atmospheric, terrestrial, and sedimentary sources, which has increased filtered total mercury concentrations in Sinclair Inlet (0.33 nanograms per liter) to concentrations greater than those of the Puget Sound (0.2 nanograms per liter). The category with the largest loading of filtered total mercury to Sinclair Inlet included diffusion of porewaters from marine sediment to the water column of Sinclair Inlet and discharge through the largest stormwater drain on the Bremerton naval complex, Bremerton, Washington. However, few data are available to estimate porewater and stormwater releases with any certainty. The release from the stormwater drain does not originate from overland flow of stormwater. Rather total mercury on soils is extracted by the chloride ions in seawater as the stormwater is drained and adjacent soils are flushed with seawater by tidal pumping. Filtered total mercury released by an unknown freshwater mechanism also was observed in the stormwater flowing through this drain.

Direct atmospheric deposition on the Sinclair Inlet, freshwater discharge from creek and stormwater basins draining into Sinclair Inlet, and saline discharges from the dry dock sumps of the naval complex are included in the next largest loading category of sources of filtered total mercury. Individual discharges from a municipal wastewater treatment plant and from the industrial steam plant of the naval complex constituted the loading category with the third largest loadings. Stormwater discharge from the shipyard portion of the naval complex and groundwater discharge from the base are included in the loading category with the smallest loading of filtered total mercury.

Presently, the origins of the solids depositing to the sediment of Sinclair Inlet are uncertain, and consequently, concentrations of sediments can be qualitatively compared only to total mercury concentrations of solids suspended in the water column. Concentrations of total mercury of suspended solids from creeks, stormwater, and even wastewater effluent discharging into greater Sinclair Inlet were comparable to concentrations of solids suspended in the water column of Sinclair Inlet. Concentrations of total mercury of suspended solids were significantly lower than those of marine bed sediment of Sinclair Inlet; these suspended solids have been shown to settle in Sinclair Inlet. The settling of suspended solids in the greater Sinclair Inlet and in Operable Unit B Marine of the naval complex likely will result in lower concentrations of total mercury in sediments. Such a decrease in total mercury concentrations was observed in the sediment of Operable Unit B Marine in 2010. However, total mercury concentrations of solids discharged from several sources from the Bremerton naval complex were higher than concentrations in sediment collected from Operable Unit B Marine. The combined loading of solids from these sources is small compared to the amount of solids depositing in OU B Marine. However, total mercury concentration in sediment collected at a monitoring station just offshore one of these sources, the largest stormwater drain on the Bremerton naval complex, increased considerably in 2010.

Low methylmercury concentrations were detected in groundwater, stormwater, and effluents discharged from the Bremerton naval complex. The highest methylmercury concentrations were measured in the porewaters of highly reducing marine sediment in greater Sinclair Inlet. The marine sediment collected off the largest stormwater drain contained low concentrations of methylmercury in porewater because these sediments were not highly reducing.

Introduction

In the 1980s, the sediment of Sinclair Inlet was known to have elevated concentrations of a number of elements and organic compounds (Malins and others, 1982). A remedial investigation of the marine waters off the Bremerton naval complex (BNC), Bremerton, Washington, was completed under the Comprehensive Environmental Response, Compensation, and Liability Act (CERCLA) in 1996 (U.S. Navy, 2002) and the final Record of Decision (U.S. Environmental Protection Agency, 2000) was issued in 2000. The remediation options included isolating a considerable volume of contaminated sediment from interactions with the benthic food web by placing dredge spoils from navigational and cleanup dredging in a covered confined aquatic disposal pit created in 2001. The primary objective of the marine sediment cleanup was to address the potential risk to humans, particularly those engaged in a subsistence lifestyle, from consumption of bottom-dwelling fish with tissue containing elevated concentrations of polychlorinated biphenyls (PCBs) (U.S. Navy, 2002). Three pathways were identified as having the capability to transport chemicals from terrestrial landscape of the BNC to the marine environment, thus have the potential to recontaminate recently remediated marine sediment. The pathways included discharges from dry dock, groundwater, and stormwater system facilities that handle surface-water runoff.

As lead agency for environmental cleanup of the BNC, the U.S. Navy completed a second 5-year review of the remedial actions for marine sediment within the boundary of the BNC (U.S. Navy, 2008a), pursuant to Section 121(c) of CERCLA and the National Oil and Hazardous Substances Pollution Contingency Plan (40 Code of Federal Regulations Part 300). One issue highlighted in the second 5-year review was, "There is insufficient information to determine whether the remedial action taken at OU [Operable Unit] B Marine with respect to mercury in sediment is protective of ingestion of rockfish by subsistence finfishers," (U.S Navy, 2008a, p. 5). Recommendations and follow-up actions in the 5-year review were:

- Revisit Remedial Investigation/Feasibility Study (RI/FS) groundwater-to-surface-water transport evaluations considering total mercury concentrations in two long-term monitoring wells,

- Perform trend analyses and assess functionality and protectiveness of remedy for marine sediment, and

- Collect additional information necessary to perform a risk evaluation and reach conclusions regarding the protectiveness of the remedy (U.S. Navy, 2002) with respect to total mercury concentrations in Sinclair Inlet sediment and fish tissue.

Since 2007, the U.S. Geological Survey (USGS) and the U.S. Navy have entered into several multiyear interagency agreements, the Watershed Project and the Methylation and Bioaccumulation Project (http://wa.water.usgs.gov/projects/sinclair/). The objective of the Watershed Project is to estimate the magnitudes of the predominant sources of total mercury to Sinclair Inlet, including those from the BNC. The objectives of the companion Methylation and Bioaccumulation Project is to evaluate the transformation of mercury to a bioavailable form in Sinclair Inlet and to assess the effect of the sources and transformation processes on the mercury burden in marine organisms and sediment. In this report, total mercury (THg) refers to all chemical forms of mercury (including methylmercury [MHg]), and not to an unfiltered water sample. A listing of published documents on mercury in Sinclair Inlet is available at http://wa.water.usgs.gov/projects/sinclair/publications htm.

Purpose and Scope of Watershed Project

The Watershed Project focused on the first objective of estimating the magnitudes of the predominant sources of THg to Sinclair Inlet. To accomplish this objective, the USGS: (1) examined the status of THg in the sediment, water, and biota of Sinclair Inlet using data available at the beginning of the study (2007); (2) assessed the sources of THg to Sinclair Inlet; (3) generated new data used to evaluate sources of THg from permitted sources and groundwater flowing from the BNC into Sinclair Inlet; and (4) made of the first measurements of MHg in the Sinclair Inlet basin.

Data for THg in sediment, water, and biota of Sinclair Inlet from other sources available at the beginning of the project were reported in Paulson and others (2010). This report describes the assessment of sources of THg to Sinclair Inlet using new data and presents the results from the first survey and measurements of MHg concentrations in the aqueous phase in sources of water discharging to Sinclair Inlet. Sources of THg from BNC groundwater, stormwater drains, and industrial sources were assessed using the new data. Where possible, the sources of THg in the aqueous and particulate forms from the BNC were compared to other sources of THg into Sinclair Inlet. The sources of THg from the BNC were compared to other sources including advection from Puget Sound, direct precipitation on the surface of Sinclair Inlet, creeks draining into Sinclair Inlet, and diffusive flux of aqueous THg from Sinclair Inlet sediment. Not all sources of mercury were examined in the same detail; therefore, the overall conclusions are limited to ranking the overall effect of the various sources that were examined. Where appropriate, this report draws on new mercury data from the water and sediment samples collected as part of the companion Methylation and Bioaccumulation Project.

Study Area

Sinclair Inlet

Sinclair Inlet is a shallow (maximum depth of 20 m) embayment on the west side of the Puget Sound lowland (fig. 1). The major axis of the inlet is aligned about 65 degrees clockwise of north (along 65 degrees true). The Puget Sound lowland is a long, northward-trending structural depression situated. Most of the Puget Sound lowland physiographic province is mantled with thick glacial and postglacial deposits.

Sinclair Inlet is adjacent to Dyes Inlet, another shallow embayment (fig. 1). The Dyes Inlet-Sinclair Inlet system is hydraulically complex not only because of the geometry of the connection, but because Bainbridge Island constricts this connection between the Dyes Inlet-Sinclair Inlet system and central Puget Sound. The Dyes Inlet-Sinclair Inlet system is connected to central Puget Sound through a convergence zone in which water flows through Port Madison, and Agate and Port Orchard Passages on the north side of Bainbridge Island, and through Rich Passage on the south side of Bainbridge Island. Rich Passage shallows to 20 m; the maximum depth in Agate Passage is 6 m. The shallowness of these passages results in extensive vertical mixing of the incoming tidal water. Tides in Puget Sound are mixed diurnally and have a maximum tidal range of about 5 m relative to a maximum depth of about 20 m for Sinclair Inlet. The relative proportion of tidal volumes through Port Orchard and Rich Passages is unknown. Because the increase in volume of water in Dyes Inlet between low tide and high tide is about three times that of Sinclair Inlet, tidal currents in Port Washington Narrows, which connects Dyes Inlet to Sinclair Inlet, often lag the tidal currents of Sinclair Inlet (Wang and Richter, 1999).

For the purpose of mass balance calculations in this report, a specific volume of water (a "box") was defined. The surfaces of the box of the Sinclair Inlet model for the mass balances include the surface of marine sediment on the seabed upward to mean sea level, the sea surface that intersects the sediment surface at the shore, and a vertical plane at the outer boundary of Sinclair Inlet—defined in this study as the seaward side of a cable area extending from the Bremerton dock of the Washington State Ferry System to the pointed shoreline near Annapolis Creek (fig. 2). Using the North American Vertical Datum of 1988, this definition of the mass balance box yields a volume of 75.7×10^6 m^3, a surface area of 8.37 km^2, and a cross-sectional area of 27,050 m^2 at its entrance.

Sinclair Inlet has been characterized as "a tidally dominated, non-stratified, saline body of water" (U.S. Navy, 1992, sec. 2, p. 6) because of the vertical mixing and the relatively small inflow of freshwater. Gartner and others (1998) determined that Sinclair Inlet was non-stratified and isothermal in August 1994. During the wet season in March 1994 when surface runoff would be near its maximum,

stratification was weak with the salinity difference between upper and lower layer less than 1 in 30 units on the Practical Salinity Scale (PSS). Other studies (Albertson and others, 1995; Katz and others, 2004) have shown that Sinclair Inlet is stratified under certain conditions. Gartner and others (1998) determined that typical current speeds were 5–10 cm/s. Wind forcing caused residual currents (time-averaged currents filtered with a 35-hour low-pass filter) in the bottom layer to be in the opposite direction of the surface layer currents and wind direction. Unlike systems dominated by estuarine circulation, residual currents in Sinclair Inlet were similar in magnitude to the tidal currents.

Bremerton Naval Complex

The Bremerton naval complex covers about 2 km^2 on the north shore of Sinclair Inlet in Bremerton, Washington (fig. 3) and houses two Navy commands: Puget Sound Naval Shipyard and Intermediate Maintenance Facility (PSNS&IMF), Bremerton site, and Naval Base Kitsap at Bremerton (NBK Bremerton). The primary role of PSNS&IMF (1.5 km^2), a fenced, high-security area, is to provide overhaul, maintenance, conversion, refueling, defueling, and repair services to the naval fleet. The primary role of NBK Bremerton (0.4 km^2) is to serve as a deep-draft home port for aircraft carriers and supply ships.

For the purposes of environmental remediation, the BNC was divided into Operable Units (OU): OU A, OU B Marine and Terrestrial, OU C, OU D, and OU NSC (Naval Supply Center) (fig. 3). In this report, only data previously collected within the OU A, OU B Terrestrial and Marine, and OU NSC is addressed; the greater Sinclair Inlet is defined as the area in the box of the Sinclair Inlet model (see above) that is outside of OU B Marine.

For this investigation, the BNC was divided into three areas defined by the flow of groundwater: (1) the Zone of Direct Discharge, (2) the Vicinity of Site 2, and (3) the Capture Zone of the Sumps associated with the dry docks (fig. 4). The Zone of Direct Discharge of groundwater to Sinclair Inlet is in the western part of BNC and includes OU A and the western part of NBK Bremerton (fig. 3). The Vicinity of Site 2 (fig. 5) is an area most affected by mercury soil contamination (URS Consultants, Inc., 1991) and includes the largest stormwater drain (PSNS015) on the base that passes through Site 2 and drains 0.41 km^2 of the NBK Bremerton (fig. 3). The area of NBK Bremerton served by the stormwater drain contains facilities for parking, housing, shopping, recreation, and dining for military personnel and their families. Groundwater from the largest area of BNC, which includes OU NSC and the PSNS&IMF area of OU B Terrestrial, is captured by the sumps of six dry docks that are located 13 m below ground surface (Prych, 1997). A seawall along most of the shore in this area facilitates the capture of groundwater by the sumps.

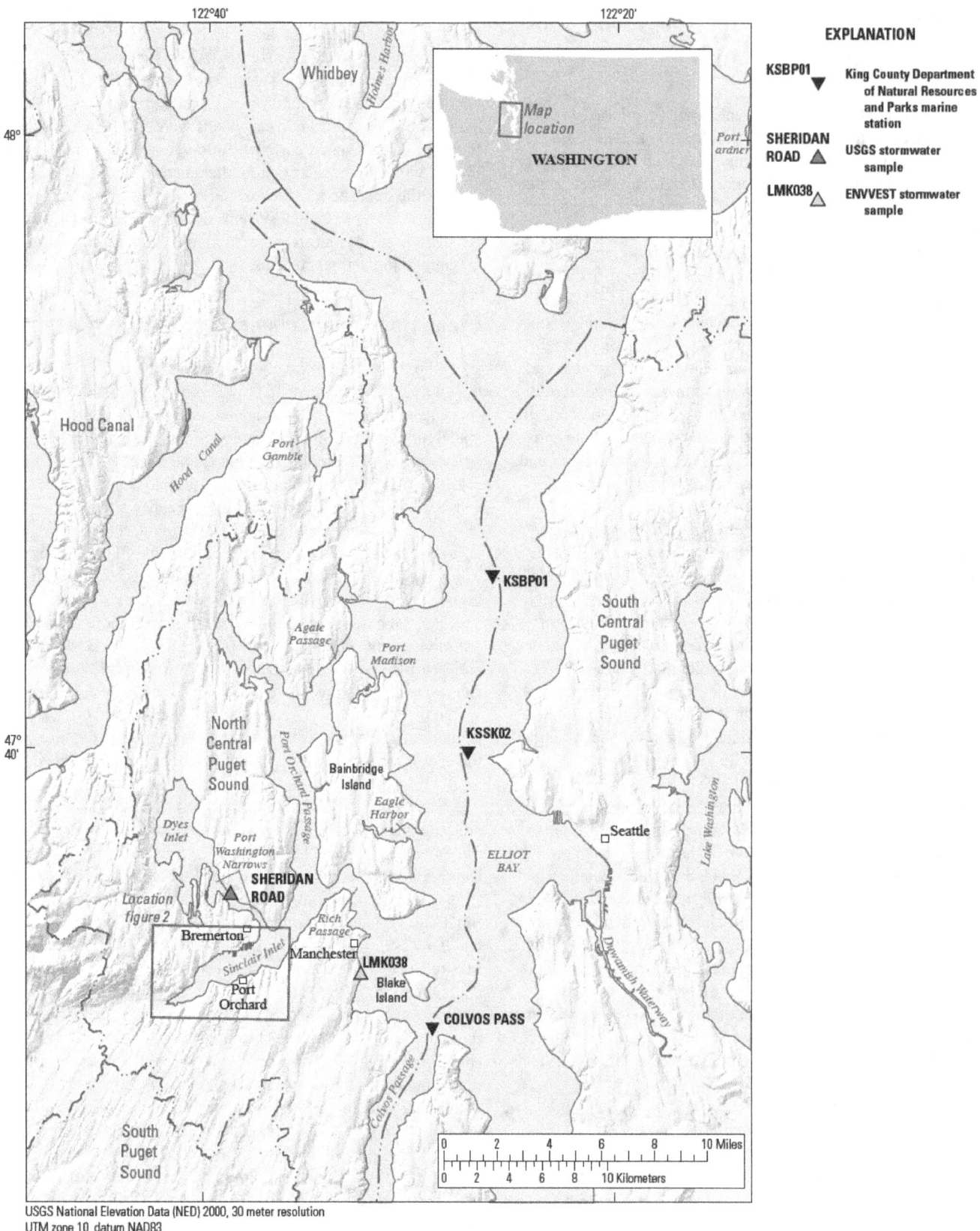

USGS National Elevation Data (NED) 2000, 30 meter resolution
UTM zone 10, datum NAD83

Figure 1.   Locations of Sinclair and Dyes Inlets, selected marine water-quality stations in central Puget Sound, and stormwater stations outside of Sinclair Inlet, Washington.

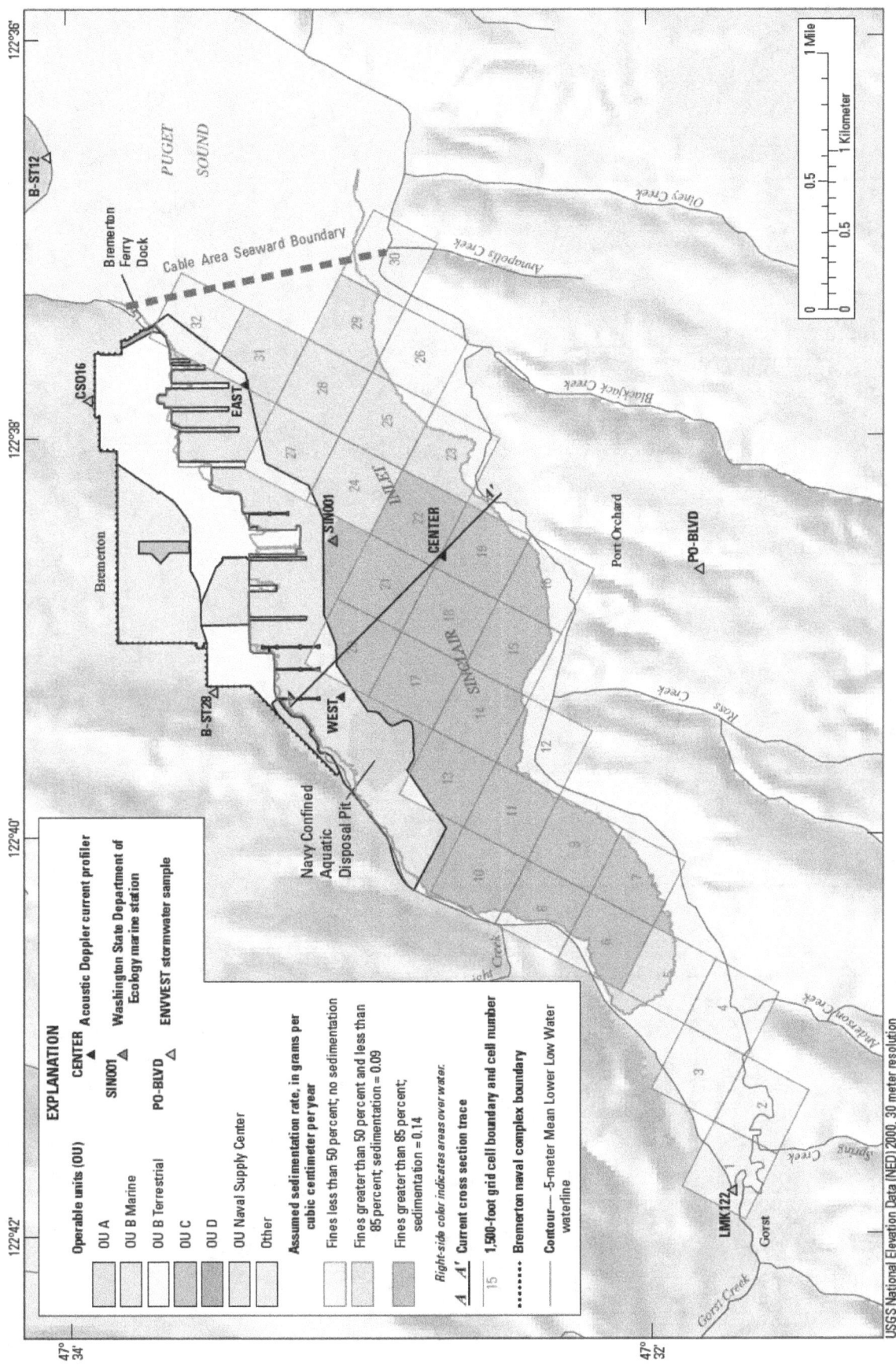

USGS National Elevation Data (NED) 2000, 30 meter resolution
UTM zone 10, datum NAD83

Figure 2.   Location of the U.S. Navy Long-Term Monitoring Program cells in the sediment-sampling grid, stormwater sites sampled by ENVironmental inVESTment project, marine station of the Washington Department of Ecology and USGS acoustic Doppler current profiler site, Sinclair Inlet, Kitsap County, Washington.

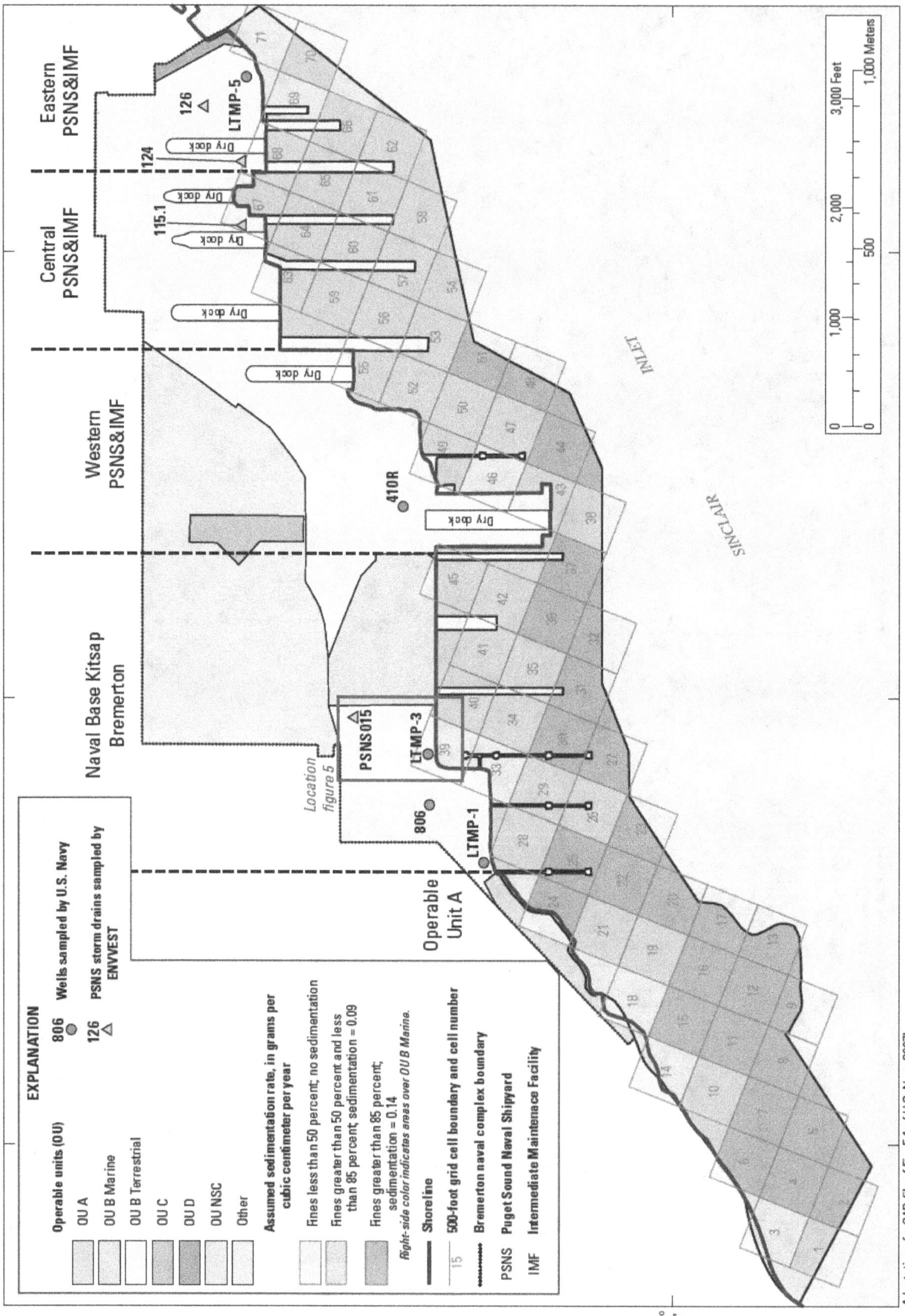

Adaptation of a CAD file of Fig. 5.1 of U.S. Navy 2007b

Figure 3.   Bremerton naval complex along the northern shore of Sinclair Inlet, Kitsap County, Washington.

Zone of Direct Discharge

OU A consists of approximately 0.049 km² of manmade land and the adjacent marine environment of Sinclair Inlet. It is located southwest of the main Puget Sound Naval Shipyard operations. The site is relatively level except for the riprap seawall, and most of its surface is paved. OU A is currently used as a parking lot for shipyard and deployed personnel and has been temporarily used as a staging area for dredge spoils. In the past, it was used as a helicopter pad and as a disposal area for industrial wastes associated with shipbuilding and decommissioning. Most of the land at OU A is composed of industrial fill to a depth of 2.1 to 10.7 m below the current ground surface. The fill increases in thickness toward the shore, where it is overlain with a layer of riprap stone. This industrial fill is composed of sandblast grit, scrap metal, brick, glass, wood, and other debris. Several shallow pits were established near the former helicopter pad and used for disposal of liquid wastes. OU A includes 0.003 km² of impervious area and is served by two stormwater outfalls (U.S. Navy, 1995a).

The Zone of Direct Discharge also includes some area in OU B Terrestrial including a steam plant, which is a permitted point source for discharge. At the time of 2008 sampling, municipal water was demineralized using ion exchange water softening technology. The effluent from the steam plant contained demineralizer regeneration and corrosive drain wastes that were neutralized and clarified through sand filters and gravity fed to the effluent wet well. Final pH adjustments were made in the wet well, if necessary, and the water was discharged into Sinclair Inlet. After USGS sampling ended in 2008, a new reverse osmosis system replaced the ion exchange system in 2010. This upgrade in technology likely changed the loadings of mercury described in this report.

A concrete tank used to neutralize acids, bases, and spent electroplating solutions near replacement well OUBT-715R (fig. 4) was observed to be degrading to the extent that a metal reinforcement bar was exposed. This observation prompted shipyard personnel to arrange alternative liquid waste disposal methods and to take the tank out of service in 1983.

Vicinity of Site 2

Site 2 (fig. 5) of the Initial Assessment Study (IAS) (URS Consultants, Inc., 1991) is just west of the OU NSC and approximately 0.02 km². Historically, Site 2 contained numerous wooden and concrete block structures including a garbage can cleaning facility and a storage area for PCB waste and off-line transformers. In the 1980s, two dark-stained soil spots were removed during a time-critical removal action near former building 399. Soils (to 12 m below ground) from monitoring wells (MW01–MW05) and boreholes (H-101–H-115) were contaminated with THg

(U.S. Navy, 1992). THg concentrations ranged from 6.6 to 31 mg/kg, with a median of 17.5 mg/kg for 42 soils samples (table A1). Subsequently, the buildings were demolished; some near-surface soils contaminated with mercury, lead, and other substances were removed; and a hazardous-and-flammable materials warehouse and associated parking areas were constructed on the site from 1994 to 1996. This area is serviced by the largest stormwater drain system (PSNS015) on the base and extends about 1 km northward and about 600 m eastward, drains 0.19 km² of impervious area, and is the only stormwater drain at BNC that services a large extent of pervious areas (0.21 km²), such as recreational ball fields. The 1.2-m diameter stormwater drain pipe passes through the seawall, then drops vertically and extends about 30-m horizontally into Sinclair Inlet between a pier and a mooring. At the seawall, the pipe is situated between +0.61 m and 0.61 m (relative to mean lower low water [MLLW]) and much of the contributing stormwater drain system is tidally affected (Bruce Beckwith, Puget Sound Naval Shipyard, written commun., 2008).

Capture Zone of Sumps

Groundwater in OU NSC and PSNS&IMF of OU B Terrestrial is captured by dry dock sumps (fig. 4). PSNS&IMF (fig. 3) contains six dry docks and the 0.75 km² of impervious area is serviced by 58 stormwater drain systems. To keep the dry docks dry when in use, sump wells are pumped intermittently between 15 to 60 minute cycles. The time-averaged discharge rates presented in this report include times when pumps of the sump wells are on and off. The depths of the sump wells for dry docks are as deep as 13 m below ground surface. The sump wells for five eastern dry docks (Dry Docks 1–5) are interconnected, and sump water was pumped through either Pump 4 or Pump 5 at an average annual discharge rate of 0.157 m³/s (Bob Johnston, U.S. Navy, written commun., 2012). When large ships are being repaired in the largest dry dock (Dry Dock 6), the cooling water from Sinclair Inlet pumped through ships also flows through the sump well at a discharge rate as much as 0.3 m³/s. When Dry Dock 6 is empty or when the ship being repaired does not discharge cooling water, the sump well is pumped at an annual discharge rate of about 0.20 m³/s (Bob Johnston, U.S. Navy, written commun., 2012). This pumping of the dry dock sump wells results in a zone of groundwater-level depression that captures most of the groundwater that would otherwise flow to Sinclair Inlet from the area between the eastern boundary of PSNS&IMF and near the western boundary of OU NSC (Prych, 1997). Additionally, significant amounts of seawater from Sinclair Inlet leak into the dry docks directly or through sediments below the sediment-water interface outside of the dry docks.

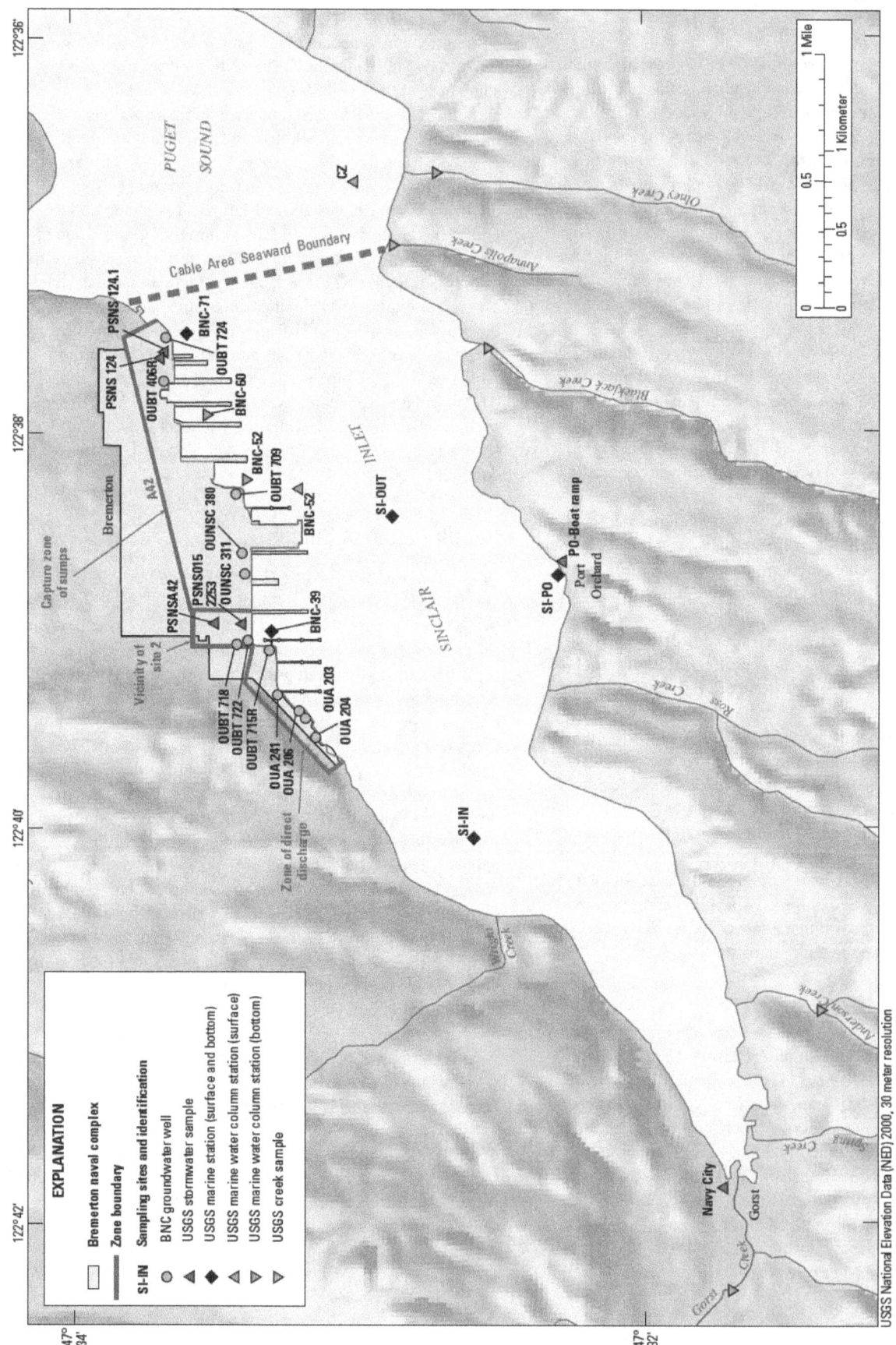

USGS National Elevation Data (NED) 2000, 30 meter resolution
UTM zone 10, datum NAD83

Figure 4. Locations of well and stormwater drain sampling sites in the Bremerton naval complex, stream and stormwater sampling sites, and marine water column stations sampled by USGS between December 2007 and March 2010, Sinclair Inlet, Kitsap County, Washington.

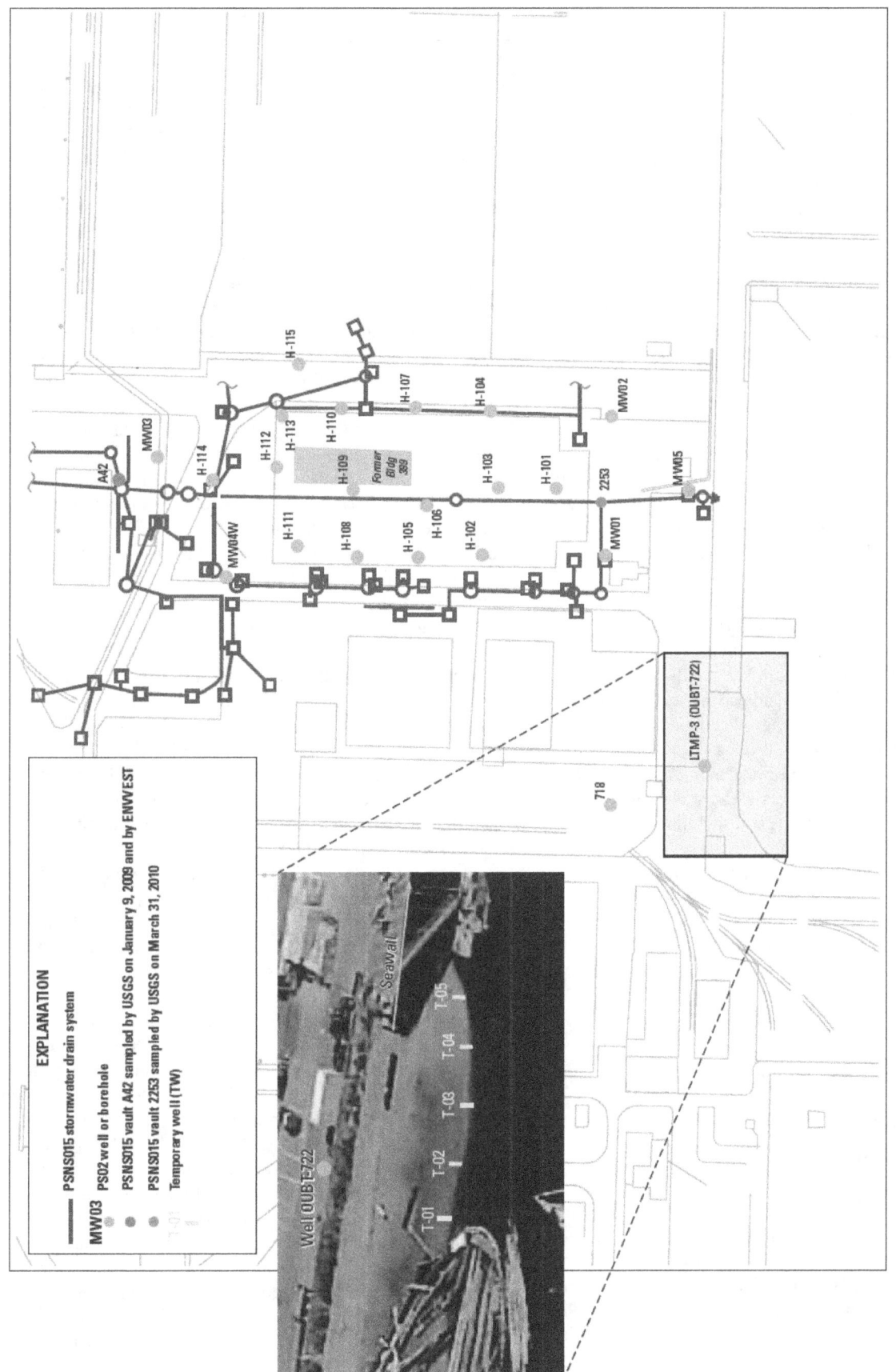

Figure 5. Vicinity Site 2 of the Initial Assessment Study including boreholes (labeled H) and decommissioned monitoring wells (labeled MW), Operable Unit B Terrestrial (OUBT) wells OUBT-722 and 718, and vaults of the PSNS015 storm water system sampled by U.S. Geological Survey, Bremerton naval complex, Bremerton, Washington, 2009–10.

OU NSC is 0.11 km^2 (fig. 3) and formerly contained the Defense Reutilization Marketing Office that recycled materials and contained an acid drain pit that was removed in 1995. Much of western OU NSC consists of former disposal sites and shoreline fill areas used for leveling and extending PSNS boundaries. The material used as fill varied with the location, but included oily sludge, automobile scrap, construction debris, shipyard debris, spent abrasive grit ("blaster sand" and copper slag), and other materials. Potential contaminants include PCBs, heavy metals, organics, and organotins (U.S. Navy, 1995b). Impervious surfaces cover most of OU NSC that is served by 16 stormwater outfalls. Most of the groundwater in OU NSC is captured by the zone of depression of the drainage systems of the six dry docks.

Numerous waterfront areas in OU B Terrestrial in the PSNS&IMF formerly served as disposal sites and shoreline fill areas used for leveling and extending PSNS. These areas received fill material that included oily sludge, automobile scrap, construction debris, shipyard debris, spent abrasive grit, and other materials (U.S. Navy, 1992). Concentrations of THg in soils from these fill areas of PSNS&IMF were three times greater than background (U.S. Navy, 2002), but did not approach the levels measured in former Site 2 of NBK Bremerton (U.S. Navy, 1992).

About 0.012 km^2 along the shoreline in the center of western PSNS&IMF (fig. 3) was used as a fill area from 1960 to 1974 and contains about 53,500 m^3 of fill ranging in thickness from 11.3 to 15.2 m. Fill materials included construction debris, rubble, spent abrasive grit, and dredged sediment. In 1998, this area was paved and the Defense Reutilization Marketing Office was relocated to this area from the OU NSC. As part of the remedial action for OU B Marine in 2000 and 2001, the shoreline perimeter was stabilized. This site also is part of the disposal site and shoreline fill area in central PSNS&IMF.

The area near the eastern boundary of PSNS&IMF near well LTMP-5 (fig. 3) probably is filled with same types of material (construction debris, rubble, spent abrasive grit, and dredged sediment). This site is covered with gravel and some asphalt pavement, and currently is used for bulk materials storage.

History of Remediation and Environmental Investigations Related to Mercury

A synthesis of data on THg concentrations in sediment throughout Puget Sound indicated that concentrations were higher in OU B Marine between two waterfront piers off central PSNS&IMF than in other urban areas of Puget Sound (Evans-Hamilton, Inc. and D.R. Systems, Inc., 1987). In 1989, personnel from the State of Washington Puget Sound Ambient Monitoring Program began monitoring the marine waters and sediment of Puget Sound. The highest concentrations of THg and PCBs of all the long-term sediment monitoring sites were detected in samples collected from Sinclair Inlet during the first Puget Sound-wide sampling (Tetra Tech, Inc., 1990). During the late 1990s, dredging of OU B marine and greater Sinclair Inlet for navigational purposes for home port expansion of the BNC was proposed. Screening of marine sediment to be dredged identified a considerable volume of sediment determined to be unsuitable for open-water disposal, because of elevated contaminant concentration (U.S. Navy, 1999). A confined aquatic disposal pit (fig. 2) was developed in 2000 for disposal of dredge spoils from home port expansion and contaminated sediment for CERCLA purposes. Even after the navigational and CERCLA dredging, the level of THg contamination in Sinclair Inlet was of the same magnitude as reported for sediment from Bellingham Bay associated with the chlor-alkali plant and sediment from Commencement Bay associated with the Ruston copper smelter (Paulson and others, 2010). Long-term monitoring of THg sediment concentrations continues at 1 site in Sinclair Inlet by the State of Washington and at 32 sites in greater Sinclair Inlet and at 71 sites within OU B Marine by the U.S. Navy, as part of the monitoring plan outlined in the Record of Decision (ROD).

During 1990–95, a series of investigations related to BNC contamination of groundwater, surface water, and dry dock discharges included (1) the IAS (1990); (2) a Site Investigation (1990–91); and (3) remedial investigation/ feasibility studies (RI/FS) at OU A (1992–94), OU NSC (1993–94), and OU B Terrestrial (1993–95). The U.S. Navy and its contractors analyzed water samples for THg collected from numerous monitoring wells in the OU A, OU NSC, and OU B Terrestrial. Water samples collected from monitoring wells in the OUs were analyzed for mercury in a Synoptic Groundwater Study between 1998 and 2003 (Grady May, U.S. Navy, written commun., 2007).

The LTMP of the BNC, which began in 2004, continues to monitor THg in water from five wells installed during the RI/FS and from five new wells installed near the shoreline at OU B. Most analyses were performed on unfiltered water, but a subset of filtered samples also was analyzed. The second five-year review for the BNC (U.S. Navy, 2008a) identified mercury contamination in marine sediments and groundwater as an on-going concern. The LTMP switched to a more sensitive analysis of THg in groundwater in late 2008.

The ENVironmental inVESTment (ENVVEST) project was developed between Federal, State and local partners to specifically address the development of total maximum daily loads for the Sinclair and Dyes Inlet basin and to improve the environmental quality of the adjacent Puget Sound Naval Shipyard; the Water Quality Implementation Plan was published in 2012 (Washington State Department of Ecology, 2012a). Scientists from the ENVVEST project assessed the sources and effects of fecal coliform (FC) pollution within the

Sinclair and Dyes Inlets (May and others, 2005). ENVVEST also developed a watershed-receiving water model for FC pollution, fate, and transport to support the total maximum daily load assessment of FC loading (Johnston and others, 2009); assessed the sources of contamination, including Hg, from creek basins and stormwater runoff as function of upstream land use and cover (Brandenberger and others, 2007; Cullinan and others, 2007); and evaluated chemical contaminant levels in demersal fish and invertebrates in Sinclair Inlet and other areas of the Puget Sound (Johnston and others, 2007).

In August 2009, the ENVVEST ambient monitoring program was initiated (Johnston and others, 2011) with the following objectives: (1) establish a baseline for assessing continuous process improvement of shipyard operations and other sources of contamination into Sinclair and Dyes Inlets; (2) provide data for validation of proposed mixing zones and model verification; (3) obtain data and information on toxicity of effluents and receiving waters for National Pollutant Discharge Elimination System (NPDES) permit requirements for the shipyard; and (4) develop procedures needed to meet ambient monitoring requirements. Additionally, Brandenberger and others (2010) evaluated the atmospheric deposition of selected metals (including Hg), polyaromatic hydrocarbons (PAHs), polybrominated diphenyl esters (PBDEs), and biomarkers to assess potential inputs of combustion-derived constituents to the Puget Sound airsheds.

Synthesis of Existing Data

Types of analyses of total mercury are presented by media of the sample. THg in an unfiltered whole water (unfiltered) sample measured by other agencies is defined as WTHg and is reported in concentrations units of nanograms per liter (ng/L). Likewise, concentrations of MHg in unfiltered water are referred to as WMHg. Total mercury in the filtrate passing through a quartz fiber filter (QFF) (0.7 μm nominal pore size) measured by USGS, the U.S. Navy, and King County Department of Natural Resources and Parks (2001) is referred to as filtered total mercury (FTHg) and is reported in concentration units of nanograms per liter. Likewise, MHg passing through a QFF is referred to as filtered methylmercury (FMHg). Projects by ENVVEST, the U.S. Navy, King County Department of Natural Resources and Parks (2001), and Brandenberger and others (2010) generally report WTHg concentrations, with some also reporting FTHg concentrations.

The total mercury of solids captured by a QFF measured by USGS is referred to as particulate total mercury (PTHg) and in concentration units of nanograms per liter (ng/L). By mass balance,

$$WTHg = FTHg + PTHg. \tag{1}$$

In this study, FTHg and PTHg generally were measured at each site. For locations where USGS and others have measured THg using modern, analytical sensitive methods, comparisons with WTHg data were made by adding FTHg concentrations and PTHg concentrations. These comparisons will be presented as part of the USGS FTHg data and allow the representativeness of the samples collected by USGS to be assessed.

The total suspended solids (TSS) concentration is a measure of solids concentrations in the part of sample collected the PTHg analysis. The total mercury concentration of suspended solids is calculated as:

$$\text{THg of suspended solids (ng/mg)} = \tag{2}$$
$$\text{PTHg (ng/L) / TSS (mg/L).}$$

By conversion, the concentrations unit of THg of suspended solids (in nanograms per milligram) is equal to the concentration unit of milligrams per kilogram (mg/kg). Total mercury concentrations of marine sediment measured by Long-Term Monitoring Program (LTMP) of U.S. Navy are referred to as STHg and reported in milligrams per kilogram.

Existing data by others relating to Puget Sound seawater, atmospheric deposition, creeks, stormwater, municipal effluents and BNC stormwater are examined for a number of mercury species in the section "Sources and Sinks of Filtered Total Mercury to Sinclair Inlet." Mercury data for groundwater outside of the BNC and diffusive fluxes of FTHg from the Sinclair Inlet sediment were not found in the literature.

Puget Sound Sources

Between June 1999 and June 2000, the King County Department of Natural Resources and Parks (2001) measured FTHg and WTHg 13 times at 3 depths at 3 stations in the main basin of Puget Sound (fig. 1): north of Bainbridge Island (KSBP01), east of Bainbridge Island (KSSK02), and at north end of Colvos Passage (COLVOS PASS), which flows north to the mixing zone with Rich Passage. For this report, mid-depth samples were selected as representative of the Puget Sound seawater that flows into Sinclair Inlet. Median WTHg concentrations for the three stations ranged from 0.33 to 0.34 ng/L. Except for one sample, FTHg concentrations were greater than the minimum detection limit of 0.1 ng/L and less than the reporting level of 0.5 ng/L. The median FTHg concentration was 0.2 ng/L.

Concentrations of WTHg at varying depths from the passages connecting Sinclair Inlet to Puget Sound were measured during the ENVVEST project (ENVironmental inVESTment Project [2006] as cited in Paulson and others, 2010). The median WTHg concentrations of bottom water in the Port Orchard Passage system and Rich Passage was 1.3 ng/L. The total suspended solids (TSS) concentrations, in a subset of 10 samples, ranged from 0.6 to 2 mg/L.

Atmospheric Deposition

Total Mercury

THg falls on the terrestrial and marine surface of the Sinclair Inlet basin as both dry particles during dry periods and as FTHg and PTHg in wet precipitation. THg on atmospheric particles, in aerosol water droplets, and as THg vapor are captured by rainfall. Brandenberger and others (2010) collected THg atmospheric deposition data during 18 two-week sampling events between September 2008 and October 2009 using bulk precipitation collectors, which collected both dry and wet deposition, at a site outside of Sinclair Inlet adjacent to Rich Passage in Manchester (fig. 1). The median concentration of WTHg was 4.75 ng/L. During two sampling events in September and October 2008 during which 1.0 and 3.8 cm of rain fell, WTHg concentrations exceeded 10 ng/L. During 13 sampling events in which 0.3–7 cm of rain fell, WTHg concentrations averaged 4.2 ± 1.3 ng/L and was not correlated with rainfall amounts (p value = 0.64).

Methylmercury

Brandenberger and others (2010) measured MHg concentrations in atmospheric precipitation collectors at a site in Manchester between September 2008 and October 2009, that included three sampling events in which little or no rain fell. Concentrations of MHg in unfiltered water (WMHg) poured and rinsed from bulk atmospheric precipitation collectors were measured during 18 two-week sampling events. The WMHg concentrations in the 60 mL of rinse water for the three sampling events in which little or no rain fall, ranged from 0.0363 to 0.509 ng/L. Concentrations of WMHg during the other 15 sampling events (125 to 797 total mL of rain water collected) averaged about 0.08 ng/L and ranged from 0.0172 to 0.263 ng/L.

Sources Discharging into Greater Sinclair Inlet

From 2002 to 2005, water samples from creeks and stormwater basins draining to Sinclair Inlet and two wastewater treatment plants were collected for the ENVVEST project. Sampling event mean concentrations were obtained using flow-weighted, time-paced, or sequential grab composites (ENVironmental inVESTment Project, 2006). Concentrations of WTHg were measured using sensitive analytical methods with a detection limit of 0.17 ng/L and a reporting limit of 0.5 ng/L for THg in unfiltered samples (ENVironmental inVESTment Project, 2006). Quality assurance data are in tables A2–A5.

Creeks Discharging to Greater Sinclair Inlet

From 2002 to 2005, four creeks (Blackjack, Anderson, Annapolis, and Gorst Creeks; fig. 2) in the Sinclair Inlet basin and many other creek basins outside of the Sinclair Inlet basin, including Olney Creek just outside the seaward boundary of the model box at the cable area (fig. 2), were sampled (ENVironmental inVESTment Project, 2006). Samples were collected for wet season base flow (March), dry season base flow (September), and a range of storm conditions (table A6). Streamflow, concentrations of WTHg, and concentrations of various other water quality constituents were measured (ENVironmental inVESTment Project, 2006).

During dry season base flow (September 2002), when TSS concentrations were 10 mg/L or less, WTHg concentrations in water from Blackjack, Anderson, Annapolis, and Olney Creeks were less than 2 ng/L. During wet season base flow and storm events, WTHg and TSS concentrations commonly increased by about one order of magnitude. During wet season base flow, concentrations of WTHg from Blackjack Creek increased to 5 ng/L or greater. Concentrations of WTHg concentrations during storms increased to about 9 ng/L in Blackjack Creek, increased to about 12 ng/L in Anderson Creek, and increased to greater than 20 ng/L in Annapolis and Olney Creeks.

The USGS examined the WTHg concentrations in surface waters in relation to total aluminum concentrations to evaluate the partitioning of WTHg between aqueous and particulate phases. Because aqueous aluminum concentrations in neutral pH waters usually are low, high total aluminum concentrations usually are associated with high concentrations of solids in the analyzed whole water sample. A strong correlation of WTHg with total aluminum in an unfiltered surface water sample would suggest that most of the THg transported during storm events was associated with particles. Concentrations of WTHg in Blackjack, Anderson, Annapolis, Upper Gorst, and Olney Creeks were well correlated (p value < 0.05; table A7) with total aluminum concentrations, indicating much of the THg was transported by particles containing aluminum. Concentration of WTHg also were correlated with concentrations of TSS in Blackjack, Anderson, Annapolis, and Olney Creeks, with the slopes of the regression ranging from 0.064 ng/mg in Anderson Creek to 0.17 ng/mg in Annapolis Creek (fig. A1 and table A7). In contrast to the four creeks previously described, WTHg concentrations collected from Gorst Creek were not significantly correlated (p value = 0.11) with TSS concentrations.

Greater Sinclair Inlet Stormwater

Stormwater samples in the Sinclair Inlet basin (table A8) were collected from one combined sewage overflow (CSO) drain (CSO16, fig. 2) and three stormwater outfalls (LMK122 in Gorst, B-ST28 in Bremerton, and PO-BLVD in Port Orchard) that discharge into Sinclair Inlet (fig. 2) (ENVironmental inVESTment Project, 2006). For comparison, stormwater data are presented from two outfalls outside of the study area: LMK038 in Manchester (fig. 1) and B-ST12 in east Bremerton (fig. 2). Samples for WTHg analysis were collected from a composite of three grab samples during storm events. Although data on many ancillary parameters were obtained from the individual grab samples, data on concentrations of total aluminum, total organic carbon, and TSS were not always obtained on the composite sample from which the aliquot for WTHg was taken. The highest WTHg concentrations in samples collected from CSO16, B-ST28, and PO-BLVD (fig. 2) were 33.8, 25.27, and 19.52 ng/L, respectively, whereas 3 of the 11 outfall samples collected from LMK122 contained WTHg concentrations between 41.9 and 56.1 ng/L (table A8). In contrast, maximum WTHg concentrations in the B-ST12 and LMK038 outfalls outside of the study area were 44.78 and 19.23 ng/L, respectively.

WTHg in unfiltered stormwater was highly correlated with total aluminum at sites CSO16, B-ST28, PO-BLVD, and LMK038 (table A7). Concentrations of WTHg also were regressed in stormwater against the concentrations of TSS (fig. A2, table A7). The WTHg concentrations of CSO16 and outfalls LMK122, B-ST28, and PO-BLVD draining into Sinclair Inlet were well–correlated ($p < 0.05$) to TSS concentrations with slopes of 0.42, 0.46, 0.19, and 0.10 ng/mg, respectively. WTHg concentrations in outfall LMK038 in Manchester were correlated with TSS concentrations (slope of 0.15 ng/mg). In contrast, WTHg concentrations in stormwater from drain B-ST12, located just outside of the Sinclair Inlet basin, were not correlated with TSS.

Concentrations of WTHg concentrations in stormwater generally were similar to WTHg measured in nearby creek water for a given storm event. During the storm event of January 17, 2005, the WTHg concentration from Annapolis Creek was 27.31 ng/L (table A6), whereas the WTHg concentration in stormwater from nearby PO-BLVD in Port Orchard was 12.95 ng/L in the outfall, and was 41.9 ng/L in the LMK122 outfall in Gorst (table A8). During the storm event of March 19, 2005, the WTHg concentration from Olney Creek was 16.96 ng/L (table A6), whereas the WTHg concentration from the CSO16 outfall in Bremerton was 15.1 ng/L and in the B-ST28 outfall in western PSNS was 11.06 ng/L (table A8). In contrast, the WTHg concentration in the B-ST12 outfall in eastern Bremerton was only 5.25 ng/L during this storm.

Municipal Wastewater Treatment Plant Effluent

Unfiltered effluent from two wastewater treatment plants were sampled and analyzed for WTHg in support of the ENVVEST project between 2004 and 2005 (table A9). Concentrations of WTHg in four final effluents from the City of Bremerton wastewater treatment plant, which discharges into Sinclair Inlet west of OU A, ranged from 8.45 to 34.8 ng/L (ENVironmental inVESTment Project, 2006). Concentrations of WTHg in eight final effluents collected from the South Kitsap Water Reclamation Facility of West Sound Utility District wastewater treatment plant, which discharges into the convergence zone east of the cable area seaward boundary (fig. 2), ranged from 4.46 to 39.83 ng/L.

Bremerton Naval Complex

Sources of THg from the BNC include surface water that is limited to stormwater drains, groundwater discharge, and the industrial discharges from the dry dock drainage relief systems and the steam plant. THg was not detected at a reporting level of 200 ng/L in any filtered or unfiltered samples collected from BNC dry dock relief drainage systems in May 1994 (U.S. Navy, 2002).

Stormwater Discharged from Bremerton Naval Complex

Surface water was sampled only in OU NSC Terrestrial during the RI/FS process (1990s) when reporting levels were 200 ng/L. Concentrations of WTHg were determined in unfiltered samples from 11 stormwater drainage systems at the BNC (table A10) for the ENVVEST project (2002–05) using sensitive analytical methods with reporting levels of 0.17 ng/L. Ancillary constituents also were analyzed in selected samples.

Zone of Direct Discharge

Concentrations of WTHg ranging from 4.92 to 42.80 ng/L were measured in stormwater from BNC drainage systems in the zone of direct groundwater discharge area (stormwater drain systems PSNS008 and PSNS011) (table A10).

Vicinity of Site 2

Stormwater from outfall PSNS015 was collected from the vault A42 (fig. 5) in the Vicinity of Site 2. High WTHg concentrations (as much as 1,131 ng/L) were of particular note (ENVironmental inVESTment Project, 2006). WTHg concentrations were not correlated with total aluminum concentrations (table A7). Although WTHg concentrations from PSNS015 were correlated with TSS, the y-intercept was -192 ng/L, which suggests caution in interpreting this correlation.

Capture Zone of Sumps

During the RI/FS process, stormwater from OU NSC was collected and analyzed for THg using a method with a reporting level of 200 ng/L (U.S Navy, 1995b). In 1993, 2 of 11 unfiltered samples collected from stormwater drains in OU NSC contained detectable WTHg concentration ranging 230–380 ng/L with registered turbidity values greater than 200 nephelometric turbidity units (NTU). The WTHg concentration of a single sample collected from OU NSC stormwater drain in 1999 was 210 ng/L, slightly above the reporting level of 200 ng/L.

The WTHg concentrations in stormwater collected in the PSNS&IMF (stormwater drains PSNS115.1, PSNS124, and PSNS126, fig. 3) for the ENVironmental inVESTment Project (2006) ranged from 11.8 to 44.0 ng/L. WTHg concentrations greater than 10 ng/L were detected even when TSS concentrations were low (fig. A2). This suggests that a significant amount of the WTHg was in the aqueous phase in BNC stormwater, or that the content of mercury in the suspended solids in BNC stormwater was higher than the content of suspended solids discharging from creek basins. Only WTHg concentrations collected from PSNS126 were correlated to TSS with a slope of 0.46 ng/mg. Additional sampling using different methods would be required to evaluate the nature of the THg in the stormwater of the PSNS&IMF.

Groundwater in Bremerton Naval Complex

Prior to the LTMP groundwater program, FTHg and WTHg concentrations throughout BNC were measured for four studies: Initial Assessment Study (1990), Site Investigation (1990–91), Remedial Investigation/Feasibility Studies (RI/FS) (1993–95), and Synoptic Groundwater Monitoring (1998–2002). Using methods with detection limits ranging between 100 and 200 ng/L, a total of 308 measurements of FTHg concentrations were qualified as not detected or estimated, and 46 of 316 measurements of WTHg were greater than detection limits between 100 and 200 ng/L. High turbidity, high TSS concentrations, and (or) high aluminum concentrations in unfiltered water indicate that 40 of these measurements were not representative of the groundwater because of inadequate well development.

The Record of Decision provides for post-remediation monitoring to be used in five-year reviews to assess the effects of the initial and subsequent remedial actions. The LTMP examines groundwater resources of OU A, OU B-Terrestrial, and OU NSC. The LTMP continues to monitor groundwater for THg, other dissolved metals, and organic contaminants in five wells installed during the Remedial Investigation and five new LTMP wells installed in 2004 in close proximity to the shoreline. In the second 5-year review (U.S. Navy, 2008a), mercury contamination in groundwater continued to be a concern. Most of the analyses were performed on unfiltered water using a method with a reporting level of 200 ng/L. A subset of samples was analyzed in 2005 for THg concentrations in filtered and unfiltered water. In 2008, the LTMP began collecting WTHg samples using "clean hands, dirty hands" sampling techniques and measured WTHg using a low-level analytical method, which resulted in a WTHg reporting level of 1 ng/L (Dwight Leisle, U.S. Navy, written commun., 2010). During 2008 and 2009, WTHg concentrations in unfiltered groundwater from the nine wells other than LTMP-3 likely were affected by the presence of solids; TSS concentrations in six samples ranged from 6 to 65 mg/L, and turbidity values of 55 and 42 NTUs were measured during 2008 and 2009, respectively (table A11). No analyses of methylmercury were performed.

Zone of Direct Discharge

Three WTHg concentrations, which were measured using the method that the high detection limits (2004–07) in groundwater collected from well LTMP-1, were between the detection limit and the reporting level. When methods with lower detection limits were used in 2008 and 2009, groundwater from LTMP-1 contained the second highest WTHg concentrations (156 ng/L in 2008 and 56.8 ng/L in 2009) of the 10 wells sampled (table A11).

Vicinity of Site 2

In the Initial Assessment Study, the two samples judged to be analytically acceptable using quality assurance criteria had exceptionally high WTHg concentrations from Site 2 wells and were accompanied by high aluminum concentrations in unfiltered water. Although these observations provide little information about the aqueous phase of groundwater, they confirm the presence of mercury contaminated soils at Site 2. During the Synoptic Groundwater Monitoring Study by the U.S. Navy in 2002 (Grady May, U.S. Navy, written commun., 2007), WTHg was detected only in unfiltered water collected from monitoring well 806 (121 m west of well LTMP-3) (fig. 3). WTHg concentrations in groundwater collected from well 806, located west of Site 2, ranged from 200 to 500 ng/L. No field parameters or samples for total aluminum or total suspended solids were collected in 2002 and 2003 from OU B Terrestrial to assess the bias due to high turbidity.

All wells installed in Site 2 during the RI/FS were decommissioned during the remediation of the site, so the sampling during July–August 1994 during the RI/FS (U.S. Navy, 2002) provides the only salinity data for groundwater under this site. The salinity of groundwater from well PS02-MW05, located near the seawall (fig. 5), was 23.76 on the practical salinity scale, and the salinity of groundwater from well PS02-MW01 located 50 m inland from PS02-MW05 was 6.54. About 200 m inland, salinity of groundwater from well PS02-MW04W was lower than the other well from Site 2, whereas salinity of groundwater from the well (MW03) closest to the PSNS015 stormwater drain vault A42 was 5.63.

Between 2004 and 2007, 9 of the 16 samples judged to be analytically acceptable using quality assurance criteria (detection limit of 200 ng/L) measured in groundwater (U.S. Navy, 2007a) were collected from well LTMP-3 (fig. 3) and ranged from 0.98 to 6.69 µg/L (table A12). During March, May, and July 2005, FTHg concentrations from well LTMP-3 ranged between 0.21 and 0.97 µg/L. WTHg concentrations in unfiltered water from LTMP-3 (table A11) were the highest during autumn 2008 and spring 2009 sampling (3,680 and 1,190 ng/L, respectively).

Capture Zone of Sumps

Concentrations of WTHg and FTHg were not detected at a reporting level of 200 ng/L in groundwater collected between 2004 and 2007 from six of seven wells within the capture zone of dry dock sumps (U.S. Navy, 2007a). In 2004 and 2005, four unfiltered samples from well LTMP-5 (named OUBT-724 by USGS) near the eastern boundary of PSNS&IMF contained WTHg concentrations from 0.23 to 5.24 µg/L (table A12). The value of 5.24 µg/L was obtained on October 18, 2005, at a tidal stage higher than other LTMP-5 samples collected and did not seem to be affected by excess solids (5 µg/L total aluminum and 0.96 mg/L TSS). Using the more sensitive analytical methods, groundwater from LTMP-5 collected in 2008 and 2009 contained the third highest WTHg concentrations (41.6 and 27.5 ng/L, respectively). During 2008 and 2009, the median concentration in groundwater collected from the other six wells in the capture zone of sumps was 3 ng/L, with a range from 0.33 to 30.9 ng/L.

Methods for USGS Sampling

Between December 2007 and March 2010, USGS sampled various sources of water in the BNC (fig. 4) for mercury concentrations, including: (1) industrial discharges from the steam plant in the Zone of Direct Discharge and two dry dock systems in the Capture Zone of Sumps, (2) stormwater in three drain systems in the Vicinity of the Site 2 and the Capture Zone of Sumps, (3) groundwater from 12 monitoring wells from the three zones, and (4) groundwater from five intertidal piezometers in the Vicinity of Site 2 (inset in fig. 5). Sources sampled outside of the BNC (figs. 1 and 4) included: (1) surface water in five creeks, (2) stormwater in three drains, and (3) effluent from two wastewater treatment plants. Sampling and processing procedures were in accordance with the USGS National Field Manual (U.S. Geological Survey, variously dated), and analytical procedures are described in detail in Huffman and others (2012). Between December 2007 and February 2008, equipment for sampling FTHg from dry docks and groundwater was acid cleaned at room temperature at the USGS Washington Water Science Center (WAWSC) laboratory, Tacoma, Washington, as described by Lewis and Brigham (2004) and FTHg samples stored in glass bottles was analyzed by the USGS National Water Quality Laboratory (NWQL), Lakewood, Colorado. Beginning in March 2008, all equipment and perfluoroalkoxy (PFA) bottles were hot acid-cleaned at the Wisconsin Mercury Research Laboratory (WMRL), Middleton, Wisconsin. Only acid-cleaned, 0.45 µm pore-size Meissner ALpHA® CMF 0.45-442 cartridge filters and pre-baked quartz filter filters (QFF) were used to filter water for analysis of FTHg. The methods and quality control data are presented in Huffman and others (2012) and are briefly described in this report. All data are reported in Huffman and others (2012) and deemed acceptable, except where noted below.

Sampling

Creeks

Water samples from five creeks were analyzed by USGS during May and July 2008 (fig. 4). Field properties were monitored directly in the creek, and samples were collected after pumping unfiltered water from the centroid of flow through PFA tubing using a peristaltic pump equipped with a small piece of flexible C-flex™ tubing for about 5 minutes. Water for FTHg analysis was filtered through a cartridge filter. After removing the cartridge filter, raw water was collected first into a 1-L sterile polyethylene terephthalate copolyester (PETG) bottle for PTHg analysis, then into a high-density polyethylene (HDPE) bottle for TSS analysis near the end of sampling cycle. Samples for ancillary parameters (particulate organic carbon (POC), particulate nitrogen (PN), dissolved organic carbon (DOC), filtered total iron, total manganese and nutrients, and TSS were collected using the techniques described in Huffman and others (2012), except that turbidity was only measured in May. Additionally, samples for anions and cations were filtered with the Pall cartridge filter (Aqua-Prep™ 0.45 µm pore size, 79 mm diameter) and acidified in the field. The FTHg results from the July sampling are qualified and considered to be estimated because of higher than usual concentration in the associated field filtering blank (0.33 ng/L), and a high relative percent difference (RPD) of 51 percent for two field duplicate samples (Huffman and others, 2012).

Stormwater

Three stormwater samples in the BNC and one drain in Sinclair Inlet basin (Navy City) were collected during higher low tide (-0.47 m MLLW) on January 7, 2009, during the height of a major storm (0.6 and 4.0 cm of rainfall on January 6 and 7, respectively). For the BNC stormwater drains, a 30-cm long polytetrafluoroethylene (PTFE) sampling port secured to a location-specific length of 0.25 in. outside-diameter PTFE tubing was lowered into the stormwater drain vault. Water was pumped using a peristaltic pump at about 300 mL/min through a flow chamber containing multimeter while field parameters were continuously monitored. A specific conductance reading less than 100 µS/cm ensured the absence of seawater in these tidally influenced stormwater drains. After the field parameters stabilized, a turbidity measurement was made using a Hach® model 2100P portable turbidimeter. Samples for FTHg, FMHg, PTHg and TSS were collected in a 2-L PTEG bottle; POC-PN-DOC samples were collected in amber glass bottles; and samples for analyses of filtered iron, manganese, and nutrients were filtered in the field into HDPE bottles as described in Huffman and others (2012). Similarly, the stormwater sample from the Navy City (fig. 4) site was collected by suspending the sampling port in the middle of a culvert opening and pumping water with a peristaltic pump. The other two stormwater drain samples (Sheridan Road in fig. 1 and PO-Boat ramp in fig. 4) were collected directly from the end of the stormwater drain pipe during the lower low tide 1 day later during the latter stages of the storm.

Saltwater leakage from the contaminated soils of the former Site 2 (U.S. Navy, 1992) to the stormwater drain PSNS015 and out to Sinclair Inlet was examined during two ebbing tidal periods in which little rain had fallen 24 hours prior to sampling. On December 29, 2009, and March 31, 2010, grab samples for FTHg were collected from vault PSNS015-2253 (fig. 5) as previously described for sampling of BNC stormwater. A YSI Inc. datasonde and sampling port were deployed into the stormwater drain. The individual grab samples were sent to the NWQL for analysis. Each day, an equal-volume composite of each grab sample was collected in a 2-L PETG bottle in the field, processed in the WAWSC laboratory, and delivered to NWQL for FTHg analysis and delivered to WMRL for FTHg and PTHg analysis; TSS was analyzed by the WAWSC laboratory. On December 29, 2009, rain started falling 4 hours into the ebbing tide. When salinity of the drain water started to decrease just as the top of the stormwater drain became visible, operations were terminated after collecting six grab samples for FTHG and the composite sample. Because of the complication of interpreting the effect of the rain event and the discrepancy in the result of the interlaboratory comparison of FTHG between NWQL and WMRL (Huffman and others, 2012), all data except PTHg and TSS values in the composite sample were disregarded. On March 31, 2010, the tidal cycle study was repeated with a Price AA velocity meter measuring water velocity at the center of the drain pipe using the methods of Rantz and others (1982). During the 24 hours prior to sampling, 0.13 cm of rain fell at the Bremerton Airport; 3.1 cm of rain fell between the 24 and 48 hours prior to sampling (National Oceanic and Atmospheric Administration, 2012). Light rain fell during sampling with minor decreases in salinity, and 10 grab samples were collected for FTHg and the composite sample.

Municipal Effluents

Composite samples of effluent water were collected from the wastewater treatment plants of the West Sound Sewer Utility District (Port Orchard) and the City of Bremerton in May, July, and August 2008. Five times during the daytime shift, an employee of each wastewater treatment plant collected a grab sample of final effluent after chlorination in a 1-L PETG bottle. At the end of the shift, USGS personnel picked up the five bottles (chilled over ice) and transported them to the WAWSC laboratory. The effluent from the West Sound Sewer Utility District is not included in the mass balance of mercury to Sinclair Inlet because it does not discharge into Sinclair Inlet and is presented for comparison purposes only.

Bremerton Naval Complex Industrial Discharges

Water samples were collected by USGS in the effluent wet well of the steam plant in the Zone of Direct Discharge monthly between March and June 2008. Because the PTFE sampling port would not fit between the gratings at the top of the wet well, a coupling containing a 10 cm piece of open 1.3-cm diameter PFA tubing was attached to 0.64-cm PTFE tubing and samples were collected using a peristaltic pump in the same manner described for collecting BNC stormwater samples.

For the first sampling event of the drainage relief systems of Dry Dock 1–5 from Pump 4 (December 2007) and Dry Dock 6 (February 2008), a grab sample was collected near the elevation of the sump and a composite sample was collected using an automatic sampler located several levels above the sump. The grab sample was filtered through a cartridge filter into 250 mL glass bottles. After February 2008, the sample was filtered into a 500 mL PFA bottle. Because stagnation of water in the vertical riser between the drainage sump and the automatic sampler, located on the upper levels of the sump wells, could affect the biochemistry of the ambient drainage water, the composite sampler location was deemed unsuitable for continued collection of mercury samples, especially for the FMHg. Following the first sampling, all dry dock samples were collected from the sump level of the dry dock drainage relief system. Dry Docks 1–5 were sampled directly from a port on the discharge pipe of Pumps 4 or 5. Dry Dock 6 was sampled by lowering the sampling port assembly, as described for stormwater sampling, through a standpipe near the pump and collecting sump water using a peristaltic pump.

The six dry dock samples collected between December 2007 and February 2008 were analyzed by the NWQL, which has a reporting level of 10 ng/L for FTHg concentrations. Because of the failure to detect FTHg concentrations at a reporting level of 10 ng/L, industrial samples for FTHg collected between March and June 2008 were analyzed by the WMRL using a method with a reporting level of 0.04 ng/L.

The range of median concentrations of the mixed data set was calculated by setting undetectable concentrations analyzed by NWQL to zero (lower limit) and reporting 6 ng/L as estimated by NWQL (upper limit) (Helsel and Hirsch, 1992). All PTHg concentrations were measured by the WMRL.

In April 2008, sample collection for FMHg concentration from Dry Docks 1–5 and Dry Dock 6, and the steam plant included a number of ancillary parameters associated with organic carbon and reduced constituents, as well as POC-PN-DOC, filtered total iron and manganese, filtered nutrients, and unfiltered ferrous iron and sulfide.

Groundwater

Base-Wide Sampling

Twice in 2008 (January 28–February 2 and April 21–24), USGS collected water samples from 11 BNC wells (fig. 4). Four wells were in OU A (OUA-203, 204, 206 and -241) in the zone of direct discharge. Two OU B Terrestrial wells were sampled in the Vicinity of the Site 2 (OUBT-722 and OUBT-718). Five wells (OUNSC-311 and -380, and OUBT-709, -406R and -724) were sampled within the capture zone of the dry dock sumps. Groundwater from shoreline well OUNSC-311 reflects OU NSC groundwater being captured by the dry dock sumps. The nearby, deeper, well OUNSC-380 contained fresher water, which likely represents ambient regional groundwater. A well in western PSNS&IMF (OUBT-709) and well in eastern PSNS&IMF (OUBT-724) were located along the shoreline in fill material. One well sampled in central PSNS&IMF (OUBT-406R) was located near the former oil tanks of a power plant.

After removal of the well cover at a well location, any water in the well cavity around the polyvinyl chloride (PVC) casing was bailed manually and with intermittent pumping from a dedicated peristaltic pump, as needed. After a water level measurement was taken, a PTFE sample port secured to a well-specific length of 0.25-in. diameter PTFE tubing was lowered into the PVC casing to the middle of the screened casing. Using a peristaltic pump, groundwater was pumped at approximately 300 mL/min, while field parameters (dissolved oxygen, temperature, specific conductance, and pH) were continuously monitored. After at least three casing volumes of water were purged and the field parameters stabilized, including turbidity using a portable turbidimeter (Hach® model 2100P), groundwater sampling began in the sequence previously described for stormwater sampling with the exception that samples for FTHg were filtered through cartridge filters in the field.

For the first groundwater sampling event (January 28–February 2) that were analyzed by the NWQL, FTHg concentrations from three of the four OU A wells, both OU NSC wells, and three of the five OU BT wells were less than the reporting level (10 ng/L). In April, FTHg samples were collected in PFA bottles and analyzed by the WMRL (reporting level of 0.04 ng/L).

In April 2008, samples for FMHg and ancillary constituents described from BNC stormwater were collected from wells OUBT-718, -722, and -724, OUA-206, and OUNSC-380. After the well was purged and field parameters stabilized, field spectrometric measurements of sulfide and ferrous iron were made. Sampling for FTHg, FMHg, PTHg, POC-PN-DOC, filtered iron, manganese and nutrients, and TSS were processed in same manner as described in Huffman and others (2012) for the collection of stormwater samples.

Tidal Studies

On May 6 2008, filtered groundwater was sampled in well OUBT-722 at about 2-hour intervals from 2 or 3 depths: near the bottom of the screened interval, the middle of the screened interval or at the water table (whichever was lowest), and near the water table when it was above mid-screen. Samples were filtered through a cartridge filters and analyzed for FTHg concentrations at the NWQL. Salinity was calculated from measured specific conductance values in unfiltered samples pumped from the sampling port located at the bottom of the screened interval. Water levels in the well periodically were measured with a steel tape, water levels in Sinclair Inlet periodically were measured with a surveyor's level and rod, and the difference between well and Inlet water levels periodically were measured using a manometer board.

At about 1-hour intervals, during June 4–5, 2008, groundwater was sampled from well OUBT-722 just below the fluctuating water table and filtered through a cartridge filter for FTHg analysis. Data for salinity and water level also was collected similar to that described for the May 6, 2008, tidal study. Additionally, five temporary piezometers (inset in fig. 5) were installed within the intertidal zone and were sampled at about 1-hour intervals (Huffman and others, 2012). Water pumped from the piezometers was filtered through a cartridge filter and analyzed for FTHg concentration by the WMRL. The specific conductance and temperature of unfiltered samples were measured in the field. Water levels in the piezometers periodically were measured with a steel tape and the difference between piezometer and Sinclair Inlet water levels periodically were measured using a manometer board. At least one PTHg and TSS sample from each of the five piezometers (nine total samples) and at least one FMHg sample from each of the five piezometers (six total samples) was collected. The only ancillary data collected in association with these FMHg samples were salinity.

Laboratory Processing

With the exception of the wastewater-treatment plant effluents and stormwater drain samples, water samples for FTHg and FMHg were filtered in the field and acidified at the WAWSC laboratory within 6 hours with 20 mL of 6 molar hydrochloric acid per liter of water in a laminar flow hood, as described by Lewis and Brigham (2004). Raw water samples were processed for analysis of PTHg, as described by Lewis and Brigham (2004). TSS was measured gravimetrically using 0.4 μm pore-size, 47-mm diameter, Nuclepore® polycarbonate filters (Huffman and others, 2012).

After returning to the WAWSC laboratory, each of the two sets of grab samples from a specific wastewater treatment plant was composited in a 5-L Teflon bottle. Each composite sample was split into five 1-L PETG bottles for processing of samples for analyses of FTHg, PTHg, FMHg concentrations and the various ancillary analyses, as previously described for sampling of stormwater. Wastewater and stormwater collected in PETG bottles were filtered through a quartz fiber filter (QFF) held in a Savillex® PFA filtering tower assembly in the laboratory laminar flow hood. The laboratory procedures, assessment of the quality of the data, and the data are reported in Huffman and others (2012).

Mass Balance of Water, Salt, and Total Suspended Solids in Sinclair Inlet

The quantification of a THg mass balance in an estuary requires a balance of water and solids that transport THg. The balance of salt within an estuary is a means to constrain and verify the balance of water. In a dispersive dominated estuary, the transfer of a chemical constituent across the seaward boundary of the estuary (net flux) requires knowledge of the gradient of the constituent along the estuary and the dispersive properties of the estuary that are controlled by geographical features and tides. Calculating the dispersive coefficient analytically from first principles for the physically complex Sinclair Inlet estuary is beyond the scope of this project. The gradient of the constituent along the estuary model can be combined with the output of a calibrated numerical baroclinic hydrodynamic model to calculate the dispersive flux of a constituent across the boundary of an estuary. Given that there are no losses or gains in the mass of salt during estuarine mixing processes, salt is an ideal constituent to assess dispersive transport. For the balance of salt in the Sinclair Inlet, the dispersive transport of salt into Sinclair Inlet from Puget Sound is constrained by the small salinity difference of 1 or 2 between Sinclair Inlet and Puget Sound that is spread over 10 km through Rich Passage and 20 km through Port Orchard and Agate Passages. Quantifying the salt balance also is complicated by the inputs near the seaward boundary of the model box that include mixing with water from Dyes Inlet introduced through Port Washington Passage and the effluent from the West Sound Sewer Utility District.

If the estuary is partially mixed, knowledge of the net advective flow into the estuary along its main longitudinal axis can be combined with concentration data of a constituent to calculate the net flux of a constituent into and out of an estuary. The validity of any numerical estuarine model can be assessed by examining the balance of salt in the estuary, because salt is neither created nor lost from the water column in a temperate climate estuary (Knudsen, 1900). The net advective flow into Sinclair Inlet and its uncertainty are dominant factor in the mass balance of THg because of the magnitude of flows relative to other water flows containing THg.

Mass Balance of Water and Salt

The measurable salinity difference between the upper and lower layers of the Sinclair Inlet water column (table 1) suggests Sinclair Inlet can be modeled as being a partially mixed estuary. A mass balance model of a chemical constituent in a partially mixed estuary requires knowledge of (1) the net advection transport, (2) the concentrations of the constituent in the new salt water entering the estuary in the lower layer, and (3) the transport of fresher water in the upper layer leaving the estuary. Because the net advective transport of water across the seaward boundary of the estuary is the small difference between the large flooding transport and the slightly larger ebbing transport, current measurements over several lunar periods are required to reduce the uncertainty in the net transport sufficiently to be useful in mass balance modeling of a chemical constituent. As with the dispersive model, the validity of the net transport from current measurements can be assessed by examining the salt balance.

Sources of Freshwater

Sources of freshwater to Sinclair Inlet include direct rainfall on Sinclair Inlet, flow of creeks, stormwater flow, effluent discharge from wastewater treatment plants, groundwater flow, and effluents from dry docks and the steam plant on the BNC. Unless freshwater is being imported into or exported out of the basin, the sum of creek and groundwater discharge should be consistent with the total precipitation on the basin minus the losses due to evapotranspiration. The groundwater discharge directly to Sinclair Inlet from the Sinclair Inlet basin outside of the BNC is not known.

Table 1.   Differences in salinity between upper and lower layers from selected sites, Sinclair Inlet, Kitsap County, Washington.

[**Station:** SIN001, WADOEb, accessed July 12, 2012; CENTER, Gartner and others, 1998. **Median salinity values:** Sal_{UL}, salinity in upper layer; Sal_{LL}, salinity in lower layer. NA, not applicable]

Station	Month of sample collection	Number of samples	Median salinity values			1- (Sal_{UL}/Sal_{LL})	
			Upper layer (Sal_{UL})	Lower layer (Sal_{LL})	Difference	Annual average	Annual deviation
			1991–2005				
SIN001	All	142	29.30	28.91	[1]0.48	[2]0.020	[2]0.008
			Dry Season				
CENTER	September	2	29.70	29.60	0.10	NA	NA
SIN001	May to October	73	29.11	29.46	[1]0.60	NA	NA
			Wet Season				
CENTER	March	7	30.07	29.51	0.56	NA	NA
SIN001	November to April	68	28.44	29.00	[1]0.36	NA	NA

[1]The median of the salinity difference of paired near-surface and at-depth values for each sampling event.

[2]Data from 2002 to 2005 were not used because three or more months of data were missing.

Atmospheric Deposition

The 118±17 cm/yr of precipitation for water years 2002–04 (Bob Johnston, U.S. Navy, written commun., 2012; ENVironmental inVESTment Project, 2006) falling on the 8.37 km[2] of Sinclair Inlet amounted to an average freshwater input of 0.31 m³/s (table 2, fig. 6).

Freshwater Discharging to Greater Sinclair Inlet

Where possible, freshwater flow estimates were used for the period of the ENVVEST project (water years 2002–05). Although annually averaged freshwater flows are used, all sources of freshwater are presented as cubic meters per second. Annual average discharge estimates from the Sinclair Inlet creek basins (1.51 m³/s) and flow estimates from stormwater drains (0.092 m³/s) were from the output of the calibrated ENVVEST project Hydrologic Simulation Program-FORTRAN (HSPF) model (Skahill and LaHatte, 2007). The average estimated discharge of the creeks to greater Sinclair Inlet was estimated to be 0.75 and 2.11 m³/s for the dry and wet periods, respectively. Precipitation falling on the 86.3 km[2] creek and stormwater drainage basins of greater Sinclair Inlet can be transported to Sinclair Inlet, lost to evapotranspiration, infiltrated into the soils, and transported directly to Sinclair Inlet as groundwater, or be diverted to wastewater treatment plants in combined sewer-stormwater systems. The water yield (length/time) was calculated by dividing the total annual flow (length³/time) of the stream and stormwater drain basins by their respective area (length²). If all the precipitation falling on the basins was transported to Sinclair Inlet, the water yield would equal the precipitation after applying the appropriate conversion factors. The average annual discharge of 1.51 m³/s for creek basins translates into a water yield of 64 cm/yr compared the 117 cm/yr of precipitation. Thus, about one-half of the rain water falling on stream basins is lost to evapotranspiration or to groundwater discharge to Sinclair Inlet, with a minor amount being lost to combined sewer-stormwater systems. For stormwater basins, the water yield was 24 cm/yr of water, with diversion to combined sewer-stormwater systems being of greater importance than for stream basins.

The annual average flow rate of the City of Bremerton wastewater treatment plant was 0.22 m³/s, much of which is imported from outside the Sinclair Inlet drainage basin. In 2009, the City of Bremerton completed a Combined Sewer Overflow (CSO) program, which reduced CSOs by 99 percent (City of Bremerton, 2012). The separation of sanitary pipes from stormwater drains will decrease the flow from the wastewater treatment plant and increase the flows from stormwater basins. The West Sound Utility District wastewater treatment plant discharges to the mixing zone outside the boundary of the domain of this box model. Additional knowledge about mixing is required before sources added to the mixing zone can be accurately included in a mass balance model (Cokelet and Stewart, 1985).

Table 2. Freshwater discharge, freshwater yields, and loadings of suspended solids to and from Sinclair Inlet, Kitsap County, Washington.

[**Source of data:** Wet deposition, water flow of stormwater drains and creeks, Skahill and LaHatte, 2007; flow and TSS from Bremerton Waste Water Treatment Plant, Patric Coxon, City of Bremerton, written commun., 2008; 2008 steam plant flows, Bruce Beckwith, U.S. Navy, written commun., 2008. **Abbreviations:** km^2, square kilometer; m^3/s, cubic meter per second; cm/yr, centimeter per year; mg/L, milligram per year; PSNS & IMF, Puget Sound Naval Shipyard and Intermediate Maintenance Facility; PSNS015, the largest stormwater drain on the base, TW, temporary well; TSS, total supsended solids; NA, not applicable; ND, not determined; ±, plus or minus; >, greater than; >>, much greater than]

Source	Area (km^2)	Annual average discharge (m^3/s)			Water yield (cm/yr)	Total suspended sediment (mg/L)		Annual loading (metric tons per year)
		Water	Standard deviation	Freshwater		Median	Range	
Sources								
Advection from Puget Sound	NA	98		NA	NA	0.7	ND	2,100
Atmospheric								
Wet	8.37	0.31	0.04	0.31	118	NA	NA	NA
Dry	8.37	NA		NA	NA	ND	ND	ND
Watersheds								
Storm drains	12.3	0.092	0.005	0.09	24	ND	ND	ND
Creeks	74.0	[1] 1.51	0.04	1.51	64	ND	ND	ND
Bremerton Waste Water Treatment Plant [2]	NA	0.22	0.01	0.22	NA	7.5	5–13	56
Total		NA		1.82				>>56
Bremerton Naval Complex								
Zone of direct discharge	0.10							
Groundwater		0.006	0.0013	NA	NA	1.20	0.2–9.4	NA
Stormwater		0.0006	0.0001	0.0006	NA	NA	NA	NA
Steam plant		0.0024		0.0024	NA	1.4	0.73–1.45	0.09
Vicinity of site 2	[3] 0.43							
Groundwater		0.0002		NA				
PSNS015								
Freshwater		0.010	0.0007	0.010	80	222		ND
Tidal flushing		0.039		NA		2.15	1.35–2.86	2.7
Capture zone of sumps	0.99							
Stormwater		0.0083	0.0006	0.008		97	35.53–158	ND
Dry docks		0.36	0.03	[4] 0.093	NA	0.85	0.16–43.6	8
Total for Bremerton naval complex	1.52			0.115				>11
Total sources		98		2.29±0.07				>>2,170
Sources discharging to passages with Puget Sound (unknown amount of flow and solids entering Sinclair Inlet)								
West Sound Utility District [2]		0.072	0.002			15	4.7–41	32
Sinks								
Advection to Puget Sound		100				2.19	0.63–12.3	6,900
Average sediment deposition from the 1960's to 2000's.								
Bremerton naval complex								1,620
Greater Sinclair Inlet								5,570
Total sedimentation								7,190
Total sinks								14,100

[1] Based on 2.11±0.01 m^3/s for the 7-month wet season and 0.75±0.16 m^3/s for the 5-month dry season.

[2] Solids loading based on monthly average total suspended solids data.

[3] Includes shoreline area between the PSNS015 stormwater drain and the western end of the seawall.

[4] Based on a salt balance using dry dock salinity from Huffman and others (2010) and Sinclair Inlet salinity from ENVVEST, 2006.

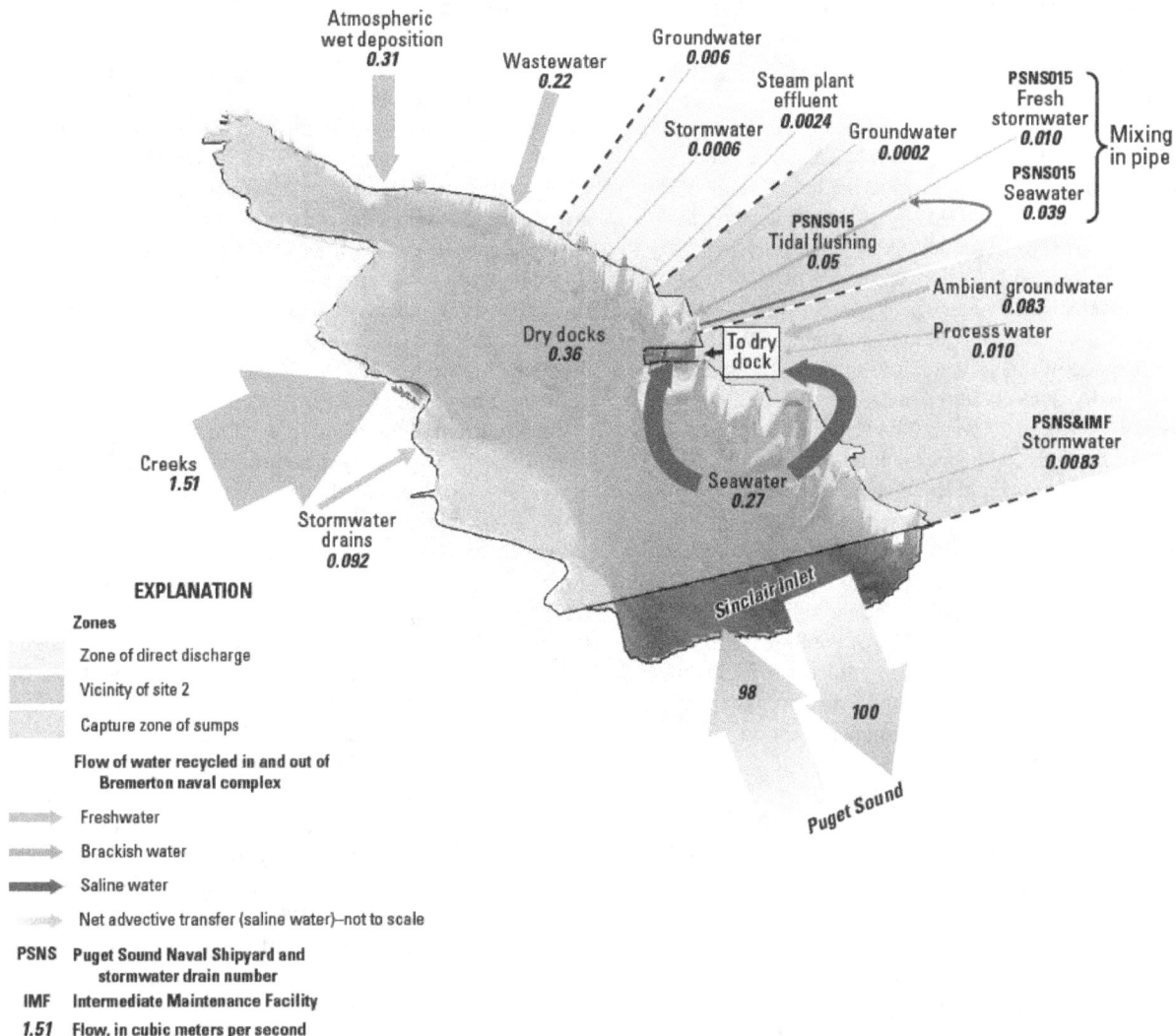

Figure 6. Flow of freshwater, seawater recycled into and out of Bremerton naval complex, and net advective transfers between Sinclair Inlet and Puget Sound, Kitsap County, Washington.

Freshwater Sources from Bremerton Naval Complex

The sources of freshwater from the BNC include discharge from groundwater, stormwater drains, the steam plant, and part of the discharge from the dry docks. The net flux of groundwater on the western portion of the BNC to Sinclair Inlet was estimated to be 0.00623 m³/s in the USGS numerical model (Prych, 1997). For comparison, the mid-range estimate of groundwater flux from OU A to Sinclair Inlet was 0.0066 m³/s using Darcy's law calculations based on measured shoreline hydraulic gradients and hydraulic conductivities, and approximated discharge areas (U.S. Navy, 1995a). The groundwater flux estimate from USGS groundwater numerical flow model of Prych (1997) is constrained by upland recharge, and the flux estimates of the U.S. Navy bracket the model estimate, so a net groundwater flux of 0.0062 m³/s was used to estimate the associated

mercury flux. For the purposes of this water balance, 0.006 m³/s is assigned to the Zone of Direct Discharge, and 0.0002 m³/s is assigned to the Vicinity of Site 2. A comprehensive error evaluation was not completed with the USGS groundwater flow model (Prych, 1997), although the author did present information to allow reasonable error estimates for the simulated direct discharge of groundwater to Sinclair Inlet from the model area. The simulated total combined (fresh and Sinclair Inlet water) discharge to the dry docks was within 4 percent of the discharge based on measurements made in 1994. It would be reasonable to infer that the error in the simulated direct discharge of groundwater to Sinclair Inlet would be similar. The relative percent difference of simulated direct discharges of groundwater to Sinclair Inlet using different prescribed boundary conditions (no-flow and specified-head boundaries) was 26 percent.

Groundwater flow simulated by a model with the correct (but unknown) boundary conditions should be between those two estimates. Combined, those two sources of error are estimated to be about 20 percent (0.0013 m³/s).

Flows from smaller BNC stormwater drains in the zone of direct discharge (0.0006 ± 0.0001 m³/s), capture zone of the sumps (0.0083 ± 0.0006 m³/s), and the large PSNS015 stormwater drain (0.010 ± 0.0007 m³/s) were taken from the ENVVEST project HSPF model (Skahill and LaHatte, 2007). In 2008, the freshwater flow from the steam plant equipped with an ion-exchange demineralizing system was 0.0024 m³/s (Bruce Beckwith, U.S. Navy, written commun., 2008).

The seepage of the seawater into the sump drainage system of the dry docks resulted in salinities of discharge water ranging between 17.7 and 24.4 (Huffman and others, 2012). Water discharged to Sinclair Inlet from the dry dock relief drainage systems includes fresh groundwater, additions of freshwater from naval operations, and seawater seeping into the sumps either directly or through the aquifer. A mass balance of salt can be used to calculate the percentage of discharge water that is fresh. Knowing the salinity of dry dock discharge (S_{DD}) and the salinity of the Sinclair Inlet (S_{SI}) seeping into the dry docks, the volume fraction of saltwater (F_{SW}) in the dry dock discharge samples that were collected was calculated according to the following equation:

$$F_{SW} = S_{DD} / S_{SI}. \qquad (3)$$

The volume fraction of freshwater (F_{FW}) can be similarly calculated by difference:

$$F_{FW} = 1 - S_{DD} / S_{SI}. \qquad (4)$$

The salinity value of 28.96 was used for S_{SI} (ENVironmental inVESTment Project, 2006). During normal operations when cooling water from a dry docked ship was not being discharged into the sump of pump 6, F_{SW} were between 61 and 84 percent for the dry docks samples, respectively. The freshwater flows were 0.032 and 0.061 m³/s from Dry Docks 1–5 and 6, respectively. Similarly, Prych (1997) calculated freshwater flows of 0.025 and 0.054 m³/s, respectively. The average annual flows of fresh process water discharged to the sumps were 0.0075 and 0.0023 m³/s from Dry Docks 1–5 and 6, respectively. (Mark Wicklein, U.S. Navy, written commun., 2012). A total of 0.27 m³/s of seawater was recycled through the dry docks.

Net Advective Transport across the Seaward Boundary

Tidal currents that slosh back and forth across a vertical plane, such as the cross section of the cable area seaward boundary of Sinclair Inlet (fig. 2), do not necessarily transport contaminants into or out of the estuary, but can transfer contaminants away from areas of elevated concentrations by dispersion. In the absence of a horizontal concentration gradient of a contaminant caused by large contaminant sources, contaminants can be moved into and out of the estuary by an advective flow. In classical estuarine circulations, heat from the sun and freshwater delivered to the upper layer of an estuary stratifies the water column of an estuary and drives advective flow out of the estuary in the fresher upper layer. The flow of salt out of the estuary in the upper layer must be replaced with inflow of saltier water in the lower layer; otherwise, the estuary would become a wide freshwater river. Because the advective flow is a small net residual current between the much larger incoming flooding and outgoing ebbing tidal currents, verification of the net residual current requires an order of magnitude greater scientific effort than only measuring tidal elevations and currents. A variety of methods are used to calculate advective flow, all of which require extensive field measurements for verification. The common approaches to calculate net advective flow across the seaward boundary are (1) deploying field instrumentation to directly measure current speeds over a time period sufficiently long enough to provide the statistical power to measure a small difference between two large numbers (Noble and others, 2006); (2) simultaneously solving the balance of water and salt by measuring freshwater flow and the salinity distribution within the estuary over several annual cycles (Cokelet and Stewart, 1985); and (3) integrating the simulated velocities over the water column derived from a baroclinic three-dimensional hydrodynamic model that has been field verified for the mass balance of salt and water (Khangaonkar and others, 2011).

The most direct approach of determining net advective flows is to measure currents over a period that is representative of the hydrologic cycle of the estuarine system. Although other transport measurements have been made in Sinclair Inlet (Katz and others, 2004), the data record for the 1994 USGS deployment (Gartner and others, 1998) is still the only data set of sufficient time span to allow calculation of net advective transport. Direct long-term measurements of currents in Sinclair Inlet were undertaken in 1994 by the USGS during 6.5 weeks in late winter and 4.5 weeks in summer using bottom-mounted acoustic Doppler current profilers (ADCP) in which resuspension of bottom sediments was emphasized. A bottom-mounted ADCP sends acoustic signals up the water column, and water velocities are determined by the Doppler shift of returning acoustic signals reflecting off particles moving with the water. Although low-pass filtered (30-hour moving averages) velocities were reported and discussed in Gartner and others (1998), net advective flows and the corresponding uncertainties for the entire 1994 current velocity data record were not reported. The raw ADCP data was obtained from USGS archives and subjected to a rigorous quality control evaluation that rejected a significant amount of near-surface data (M.A. Noble and others, U.S.

Geological Survey, written commun., 2012). In this analysis, the principle axis of the velocity was determined (65 degrees from north), and the velocity data was then transformed to along-estuary and across-estuary speeds. The integration of the along-estuary current speeds over the depth of lower layer determines the transport of water, salt, and contaminants into the estuary across a cross section of the estuary (fig. 2). Likewise, the integration of the along-estuary speeds over the depth of upper layer determines the transport out of the estuary. Because of the amount of near-surface data that was rejected, the data set for the upper layer is incomplete and the balance of water could not be verified. The uncertainties in the net along-estuary speed at each depth are large because the net advective residual current is a small difference between the much larger flooding and ebbing tides. The lag time between current measurements for the same depth over which speeds are no longer autocorrelated determines the duration of each independent measurement, and thus determines the number of independent measurements during each deployment. The across-estuary current cross section trace that intercepts the CENTER site (fig. 2) is divided into two portions with currents measured at the CENTER site representing currents in the southern half of the cross section. Because the WEST site was positioned only 0.5 km west of the cross section, currents obtained from the West site are assumed to represent the currents in the northern half of the section. The results of the ADCP deployments from these two sites and third site (EAST), not used in the transport calculation, are detailed in M.A. Noble and others, U.S. Geological Survey, written commun., 2012.

During the late winter 1994, the strongest landward current speeds in the lower layer (2.2 cm/s) were observed at mid-depth at the CENTER site (fig. 7). Integration of along-estuary current speeds over the lower layer (eight 1-m depth intervals in which the net advective current speed was significantly greater than zero) yields a transport of 102 ± 17.4 m³/s for the southern half of the cross section. In contrast, landward current speeds significantly greater than zero were observed only in four 1-m depth intervals in the lower layer at the West site with strongest net along-estuary currents being 0.9 cm/s. Integration of these net advective current speeds yields a transport of 18.9 ± 6.8 m³/s for the northern half of the cross section. Adding the transport for the two halves of the cross section yields a total transport of 121 ± 18.7 m³/s for the late winter deployment. Integrating over the entire lower layer, regardless of whether the landward along-estuary current speed was significant, increases the total transport by about 15 percent, but also increases the relative error.

In contrast to the late winter deployment, the strongest along-estuary current speeds during the summer deployment were observed at mid-depth at the West site. Integration of along-estuary current speeds over the seven 1-m depth intervals for which along-estuary current speeds

were significantly greater than zero yields a transport of 60.2 ± 9.9 m³/s for the northern half of cross section. Along-estuary current speeds at the Center site were significant only in the bottom two 1-m depth intervals and integration of this 2-m depth interval yields a transport of 12.8 ± 7.4 m³/s. For the summer deployment, the net advective transport of both halves of the cross section was 73 ± 12.4 m³/s.

There are considerable uncertainties in scaling these two observations to annual average transport values needed to apply these observations to a mass balance of THg in Sinclair Inlet. With only two short ADCP deployments, it is not clear if the late winter deployment is representative of currents in Sinclair Inlet during wet, windy weather conditions typically observed during winter and if the summer deployment is representative of currents in Sinclair Inlet during dry, calm weather conditions typically observed between July and October. Besides weighting errors in scaling these two observations to the 1994 average conditions, the issue of interannual variability in transport in the landward direction due to oceanic and climatic conditions needs to be considered. An average transport of 98 ± 23 m³/s (23 percent relative error) will be used as a long-term annual average of landward along-estuary transport in the bottom layer, but the uncertainty is high because of variable results in 1994, the unknown variability during the rest of 1994, and the unknown long-term variability.

The current and modeling studies of Wang and Richter (1999) indicate that delayed tidal currents from Dyes Inlet set up a clockwise gyre in Sinclair Inlet off the Port Orchard Passage. The above calculation of average transport assume that net residual current along the southern part of the transect A–A cross section trace (fig. 2) is similar to the net residual currents at CENTER ADCP site. If the outgoing net residual currents along the southern shoreline of Sinclair Inlet are lower than those recorded at CENTER ADCP site, then the above net transport would be overestimated. The uncertainty of transport is higher than the variation between the two seasons because of the unknown effect of this clockwise gyre on average transport into Sinclair Inlet.

The net advective transport values during these two deployments cannot be verified by balancing water in Sinclair Inlet because of the incomplete data set of current measurements in the surface layer. Even though the estuarine circulation appeared to be driven by long-term winds (M.A. Noble and others, U.S. Geological Survey, written commun., 2012), the balance of salt still needs to be maintained. The salt balance of an estuary itself has been used to provide a primary check of the net transport into and out of the estuary using a form of the Knudsen equation (Knudsen, 1900). Knudsen's equation examines the consistency between the dilution of the salt from marine water by means of freshwater and the relative flows of saltwater and freshwater.

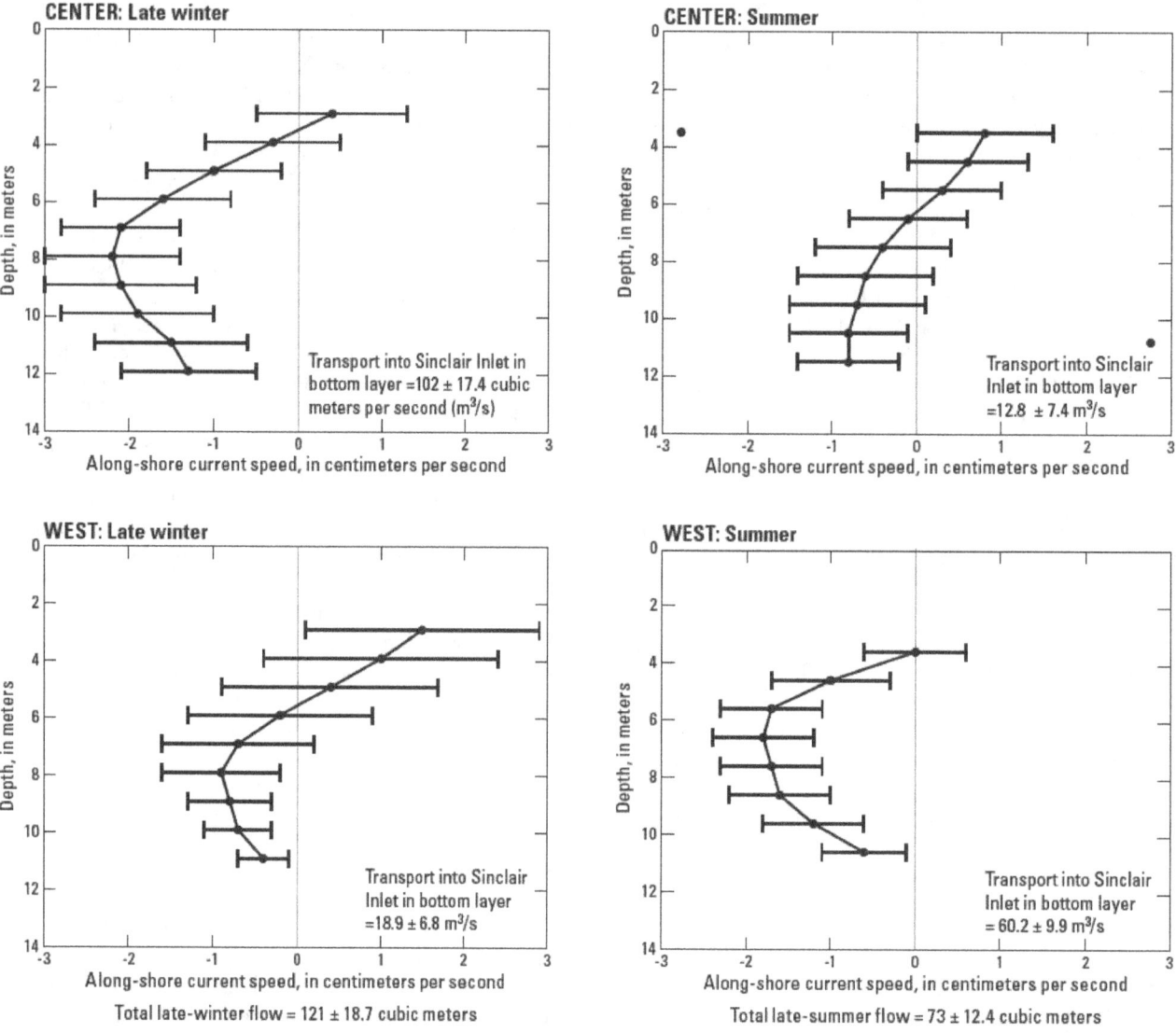

Figure 7.   Profiles of current with depth at the CENTER and WEST acoustic Doppler current profiler sites during late winter and summer 1994, Sinclair Inlet, Washington. Data from M.A. Noble and others, U.S. Geological Survey, written commun., 2012.

Consistency of the net advective flow with field observations is assessed by comparing the variables on the left side of equation 5 with the variable on the right side:

$$Q_{FW} \,/\, (Q_{SW} + Q_{FW}) \;=\; 1 - (S_{UL} \,/\, S_{LL}), \tag{5}$$

where

Q_{FW} is long-term annual average flow of freshwater into Sinclair Inlet,

Q_{SW} is the long-term annual net advective transport of seawater into Sinclair Inlet,

S_{UL} is the long-term salinity in the upper layer weighted by velocity over the depth of the upper layer, and

S_{LL} is the long-term salinity in the lower layer weighted over the depth of the lower layer.

The most rigorous application of equation 5 requires that long-term annual average S_{UL} and S_{LL} be derived by integrating salinity over the cross section weighted by the long-term along-estuary speeds as a function of depth (Cokelet and Stewart, 1985). Given the variability in profiles of the along-estuary current speeds shown in figure 7 and the fact that Q_{SW} and Q_{FW} were not measured simultaneously, a less exact application of Knudsen's equation will be undertaken as broad check of the Q_{SW}. Additionally, the salinity of bottom water must accurately reflect freshwater inputs introduced to the passages between Sinclair Inlet and Puget Sound, including the freshwater inputs of Dyes Inlet introduced by Port Washington Passage and the effluent of the West Sound Sewer Utility District introduced to the convergence zone (fig. 1).

The most easily accessible long-term salinity data set was obtained by the Marine Monitoring Program of the Washington State Department of Ecology (2012b) in which the salinities at the depth of discrete surface (nominal depth of 0.5 m) and near bottom (nominal depth of 10 m) were recorded from a conductivity temperature depth (CTD) profile taken from a float plane on a monthly basis (weather permitting) at a station just east of BNC-52 (fig. 4). The average value of the variable on the right side of equation 5 for 139 sampling periods between 1991 and 2005 was 0.018 and ranged between 0.000 (the water column being completed mixed) and 0.205 (highly stratified water column in January 1995, when the surface salinity was 22.94). The monthly values on the right side of equation 5 were averaged for 1991–2001. The 10-year average annual value of the right side of equation 5 was 0.020 ± 0.008.

The long-term variability in the freshwater input on the left side of equation 5 is dominated by the variability in the annual discharge from creek, shoreline, and stormwater drain basins. The monthly discharge simulations from a calibrated HSPF model (Skahill and LaHatte, 2007) were averaged for each year from 1993 to 2001 for the entire Sinclair Inlet drainage basin, yielding an average freshwater flow of 1.51 ± 0.04 m³/s. Adding the freshwater flows from rainfall, the wastewater treatment plant, and the dry dock relief drainage systems yield a freshwater flow of 2.29 ± 0.35 m³/s (20 percent relative error) for about the same time as the determination of the salinity difference, which encompasses the ADCP study. Applying the average transport value of 98 ± 23 m³/s for Q_{SW} to this freshwater flow, yields a value of 0.023 ± 0.006 for the left side of equation 2. Given the varying time scales and periods of the available data for freshwater flow and marine velocity and salinity, a net along-estuary advection transport of 98 m³/s seems to be valid within a factor less than 2.

The net advection of water (Q_{SW}) into Sinclair Inlet is one of the most important factors controlling the mass balance of THg in Sinclair Inlet. The magnitude of Q_{SW} dictates the amount of THg needed to be discharged to Sinclair Inlet to cause a specific increase in THg concentrations relative to THg concentrations in Puget Sound water. For instance using the best estimate of 98 m³/s for Q_{SW}, an increase of 0.1 ng/L of THg requires that about 300 g of THg be added to Sinclair Inlet to balance the mass of THg leaving Sinclair Inlet. In contrast, a discharge of about 600 g of THg would be required to balance THg for the same increase of 0.1 ng/L, if $Q_{SW} = 200$ m³/s. Thus, reducing the uncertainty in Q_{SW} is required to constrain the mass balance of THg in Sinclair Inlet.

A more exact method of verifying these transport values is to vertically integrate the average along-estuary current speeds over the cells of a baroclinic three-dimensional hydrodynamic model that accommodates wind shear and has been constrained by the balance of salt. Johnston and others (2009) developed a Curvilinear Hydrodynamics in 3 Dimensions (CH3D) model to simulate the distribution of fecal coliform in Sinclair and Dyes Inlets. The model

was calibrated using the 1994 USGS tidal current and tidal heights; the tidal currents were verified with instantaneous measurements of tidal currents made using boat mounted ADCPs (Katz and others, 2004), dye tests, and drogue tests (Wang and others, 2005). Results of tidal currents and height from the model were compared to the 67 day record of current measurements from an upward looking ADCP deployed in 1997 by the U.S. Navy Space and Naval Warfare Systems Command (SPAWAR1) near the USGS Center site (Wang and Richter, 1999). Following the addition of freshwater flows and fecal coliform loading from the watershed, the integrated model was verified by simulating short term storms (10 days) and comparing model simulation results to measured salinity and fecal coliform concentrations at selected locations within the model domain. After adjusting the model with finer grid sizes in nearshore areas to better represent freshwater mixing, the model recreated measured data with acceptable accuracy for most of the model domain (Johnston and others, 2009). All of the field components needed to constrain a hydrodynamic model have been performed between 1994 and 2009, and include long-term upward looking ADCP records (1994, 1997), stream monitoring (2002–05), wind measurements (1994, 2002–05), and extensive salinity profiles over the annual cycle (1991–2005, 2008–09). However, conducting a detailed analysis of all these components to tightly constrain the balance of water and salt within Sinclair Inlet was outside the scope of this project.

Mass Balance of Solids

In addition to balancing salt in an estuary, the mass of solids entering the water column of the estuary from air, land, or transported from Puget Sound needs to be balanced with the mass of solids leaving the water column of the estuary either by sedimentation or in the upper layer flowing to Puget Sound. Balancing solids within an estuary is much more challenging because various processes, in addition to water advection and dispersion, affect solids movement. These processes include resuspension, differential settling, flocculation, and biological growth.

Sources of Solids

Without knowing the origin of particles that settle in the sediment column, the relative contribution of the mass of a constituent from different sources in the sediment is unknown. Although sand and clay particles are essentially conserved in an estuary, the production of biological organic matter and the decay of biological detritus can introduce errors in the mass balance of solids. However, the small loss of organic matter by remineralization (6 weight percent organic matter in the sediments of Sinclair Inlet from the U.S. Navy, 2006a, 2006b, and 2008b) will contribute only a small error to the mass balance of solids in Sinclair Inlet.

Puget Sound Sources of Solids

The advection of solids into Sinclair Inlet from Puget Sound requires knowledge of the net advective transport from, and average concentrations of TSS, in Puget Sound. The coarse bed sediments in Rich and Port Orchard Passages suggest that few particles settle in these passages. The TSS concentration of 0.7 mg/L for Puget Sound (Paulson and others, 1988) was used to derive an advective transport of 2,100 metric tons per year of solids to Sinclair Inlet from Puget Sound. The higher average TSS concentration of 2 mg/L for Puget Sound collected at two stations by King County Department of Natural Resources and Parks (2001) in Puget Sound was obtained by methods thought to be biased high (Battelle Duxbury, 1998).

Solids Discharging into Greater Sinclair Inlet

Because the ENVVEST project HSPF model was not calibrated for TSS transport, the loading of TSS from creek and stormwater basins is not known. Empirical calculations that do not properly weight the high flow events, when most solids are discharged to estuaries, are likely to grossly underestimate the total loadings. The sources of solids that have not been quantified in this study include: the creek basins of Sinclair Inlet; stormwater from inside and outside of the BNC (except tidal flushing of the storm drain PSNS015); bed transport from the convergence zone; suspended and bed transport from Dyes Inlet; landslide and shoreline erosion within Sinclair Inlet and its approaches from Puget Sound (Rich and Port Orchard Passages); and dry deposition. Without the quantification of solids from the stormwater and creek basins, there is an incomplete understanding of solids sources and sinks and there is a limited comparison of the quantifiable solids loadings among a few quantifiable sources (municipal, industrial and Puget Sound sources) with quantifiable loadings among sinks (sedimentation and advection to Puget Sound).

Municipal and industrial sources of suspended solids include the City of Bremerton Wastewater Treatment Plant, the BNC steam plant, and the two dry dock discharge systems. The daily flow-weighted, composited TSS values from the treatment plant were multiplied by the daily flows from the plant to derive daily TSS loadings (Patric Coxon, City of Bremerton, written commun., 2008), which were summed and averaged for water years 2002–04 to produce an average annual TSS loading of 56±4 metric tons per year.

Solids Discharging from Bremerton Naval Complex

In 2008, USGS measured TSS (Huffman and others, 2012) in grab samples of effluent from the steam plant (number of samples [n]=4) and dry dock systems (n=11). These grab samples collected during a given time in the pumping cycle may not reflect the average conditions during the pumping cycle. Additionally, the results from the limited number of samples do not reflect the range of activities that occur throughout the year. The median TSS concentration in the steam plant effluent in 2008 was 1.4 mg/L (range: 0.47 to 1.52 mg/L), which results in an estimated load of 0.09 metric tons per year. The median TSS concentrations in samples collected from the discharge pipe of dry docks was 0.85 mg/L and ranged between 0.16 and 43.6 mg/L (table 2). The exceptionally high TSS concentration of 43.6 mg/L, compared to the next highest TSS concentration of 1.47 mg/L, was measured on June 24, 2008. At the time of the June sampling (10:30 a.m. PDT), two unusual activities were occurring in the dry dock system (Bruce Beckwith, U.S. Navy, written commun., 2008): (1) repairs were being made to the drainage system that may have stirred up accumulated sediments, and (2) concrete was being cut in one of the dry docks using cooling water on the cutting blade. If exceptionally high TSS concentrations were still present in the afternoon, it is likely that the total copper concentration in an unfiltered sample collected by the U.S Navy at the 2:30 p.m. would have exceeded the detection limit of 10 µg/L. No copper was detected in the 2:30 p.m. sample, suggesting that the morning drainage repairs, the concrete cutting operation, or both were the probable cause for the temporary high TSS in the USGS sample collected in the morning. Using a median value of 0.84 mg/L for the TSS concentration and average operating flow conditions, the loading of solids to Sinclair Inlet from the Dry Dock 1–5 relief drainage system was about 4 metric tons per year. Likewise, a median value of 0.71 mg/L from the sump for Dry Dock 6 resulted in a solids loading of 4 metric tons per year.

As described in the water budget, between 61 and 84 percent of the water discharged by the dry dock pumps is seawater from Sinclair Inlet. Thus, much of the estimated 8 metric tons per year of solids discharged from the dry dock drainage system may have originated in Sinclair Inlet rather than from naval operations or groundwater seepage. However, the portion of solids originating from Sinclair Inlet cannot be quantified because there are no data on the efficiency of particle trapping along the flowpath from Sinclair Inlet to pump system sumps. Although solids are detected in monitoring wells, it is thought that particles are efficiently trapped along the flow path to the estuary in all but the coarsest soils. Thus, no solids are assumed to originate from groundwater flow.

During each of the relatively dry-period sampling events of the PSNS015 stormwater drain during ebbing tidal cycles on December 29, 2010, and March 31, 2011, TSS samples were obtained from timed-average composite samples collected during the sampling periods. The TSS concentrations were 1.35 and 2.86 mg/L, respectively. Integrating the average TSS concentrations over the 1,770 m^3 that flowed to Sinclair Inlet during the March 31, 2010, sampling event, yields a discharge of 3.7 kg per tidal cycle. Scaling the median value up for the two tidal cycles each day over the year yields an estimated annual TSS loading of 2.7 metric tons of solids discharge from tidal flushing.

Sinks for Solids

The primary sinks for solids in Sinclair Inlet are sedimentation and advection to Puget Sound. The estimated sedimentation of solids on a dry weight basis is described in detail in appendix B and calculated in table B1. This method results in an estimated sedimentation of 1,620 metric tons per year in the OU B Marine and 5,570 metric tons per year in greater Sinclair Inlet for a total of 7,190 metric tons per year for the Sinclair Inlet model domain.

Another sink for solids is advection from Sinclair Inlet to Puget Sound. The median concentration of TSS in the upper layer at the greater Sinclair Inlet stations (table 3) between August 2008 and August 2009 was 2.19 mg/L (range: 0.63 to 12.27 mg/L). Using the estimated seaward along-estuary net advection flow of approximately 100 m^3/s out of Sinclair Inlet in the upper layer translates into a possible net export of 6,900 metric tons of solids. Because suspended solids can settle in Sinclair Inlet and particulate organic matter can decompose in the water column and at the sediment-water interface, not all the estimated 6,900 metric tons of solids transported in the upper layer of Sinclair Inlet may be leaving Sinclair Inlet. The surface layer in the convergence zone is highly influenced by the outflow of water from Sinclair Inlet and Dyes Inlet through Port Washington Narrows (fig. 1). The TSS concentrations at the convergence zone can be used to assess the loss of TSS in upper layer. The median TSS concentration in the upper layer of the convergence zone (1.91 mg/L) is slightly lower than the median concentration in upper layer of greater Sinclair Inlet (2.19 mg/L). The similarity in the TSS concentrations in greater Sinclair Inlet and the convergence zone indicates that little TSS is lost in the convergence zone and the Dyes Inlet-Sinclair Inlet system is a net exporter of TSS to the central basin of Puget Sound, presumably as smaller size solids.

Table 3.   Total suspended solids, filtered total mercury, and total mercury of suspended solids from five categories of the marine water samples from Sinclair Inlet, Kitsap County, Washington, 2008–09.

[**Multi-comparison category:** Post-anova Tukey tests. **Abbreviations:** mg/L, milligram per liter; ng/L, nanogram per liter; mg/kg, milligram per kilogram; OU, operable unit]

Sample category	Total suspended solids (mg/L)					Filtered total mercury (ng/L)				
	Number of samples	Mean	Standard deviation	Median	Range	Number of samples	Mean	Standard deviation	Median	Range
OU B marine lower layer	14	9.54	26.41	2.38	1.13–101.2	13	0.31	0.14	0.25	0.21–0.72
OU B marine upper layer	20	2.14	1.35	1.87	0.7–6.05	21	0.33	0.12	0.31	0.04–0.572
Convergence zone	12	1.88	0.63	1.91	0.72–2.95	12	0.30	0.06	0.31	0.18–0.41
Greater Sinclair Inlet lower layer	16	2.40	0.93	2.32	0.92–4.06	12	0.45	0.23	0.36	0.22–0.97
Greater Sinclair Inlet upper layer	27	3.05	2.81	2.19	0.63–12.27	28	0.39	0.16	0.33	0.22–0.87

Station category	Total mercury of solids (mg/kg)				
	Number of samples	Mean	Standard deviation	Median	Range	Multi-comparison category
OU B marine lower layer	14	0.45	0.28	0.40	0.01–1.02	A
OU B marine upper layer	22	0.23	0.13	0.18	0.07–0.56	B
Convergence zone	12	0.29	0.12	0.25	0.12–0.49	B
Greater Sinclair Inlet lower layer	16	0.32	0.15	0.37	[2]0.09–0.57	B
Greater Sinclair Inlet upper layer	29	0.23	0.17	0.17	0.066–0.72	B

[1] Does not include the concentration of targeted BNC-39 sample collected on June 2, 2009.

Rationale for Separate Mass Balances of Dissolved and Particulate Mercury in Sinclair Inlet

The data on streamflow, salinity distribution in Sinclair Inlet, and 1994 ADCP current measurements, coupled with the hydrodynamic baroclinic model, constrain the uncertainty of the net advective transport across the seaward boundary of the model domain to an extent useful for pursuing the mass balance approach for FTHg. However, the sum of the quantifiable sources of TSS from watersheds (56 metric tons per year) and the Bremerton naval complex (11 metric tons per year) amounts to less than 1 percent of the mass of solids depositing in sediments of Sinclair Inlet (7,190 metric tons per year). This discrepancy is compounded because the Sinclair Inlet appears to be a net exporter of solids to Puget Sound. Without a better understanding of the sources of solids that settle in Sinclair Inlet, efforts to balance the sources of mercury with a significant sediment sink likely will be unsuccessful.

To maximize the usefulness of the available data in understanding the environmental effects of present day sources of THg to Sinclair Inlet on THg concentrations in the water and sediment column, the THg in the water and particle phases are examined separately. This approach previously has proven useful in understanding the fate of trace metals in Puget Sound (Paulson and others, 1988) and separating dissolved and particulate mass balances of THg is justified if the rate of conversion between FTHg and PTHg is small compared to the sources and transport rates. If this conversion of THg between the aqueous and particulate phases is small, then THg can be assumed to be what estuarine scientists call "essentially conservative."

Extensive studies support this approach of developing separate mass balances for aqueous and particulate THg. In San Francisco Bay, California, the seasonal variation of FTHg concentrations were attributed to temporal changes in sources (Choe and others, 2004) and FTHg concentrations were not affected by phytoplankton bloom growth or decay (Luengen and Flegal, 2009). The seasonal variation in FTHg concentrations at four stations in Sinclair Inlet between

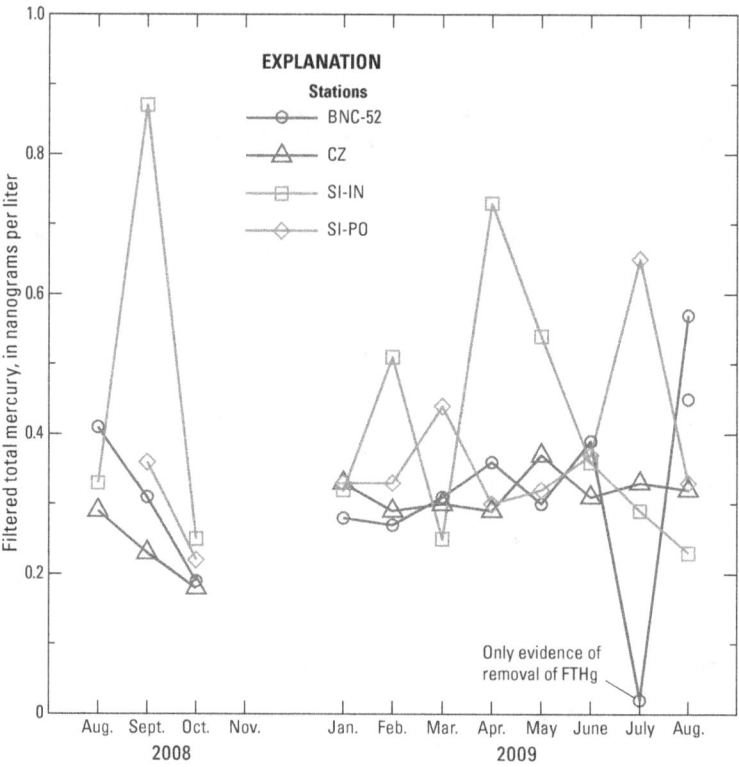

Figure 8. Concentrations of filtered total mercury in near-surface water collected from four stations in Sinclair Inlet, August 2008– August 2009.

August 2008 and August of 2009 as part of the Methylation and Bioaccumulation Project (Huffman and others, 2012) (fig. 8) were examined for similarities to the behavior of FTHg concentrations in San Francisco Bay. Occasional elevated FTHg concentrations are superimposed on a steady baseline of FTHg that is consistent among Sinclair Inlet stations. The steady baseline of FTHg in Sinclair Inlet indicates that the rate of conversion from FTHg to PTHg is small relative the transport of FTHg from Puget Sound. The only strong indication of depletion of FTHg was indicated in one sample (BNC-52) in July 2009 during a period of high productivity (chlorophyll *a* concentration of 19 µg/L, Huffman and others, 2012). The occasional elevated FTHg concentrations over the baseline probably are a temporary consequence of local sources. The decline in the baseline of FTHg during the autumn 2008 bloom period is relatively small and could be a result of a small conversion from FTHg to PTHg or decreasing inputs of FTHg.

Sources and Sinks of Filtered Total Mercury to Sinclair Inlet

FTHg discharged from non-BNC watershed originates primarily from precipitation, leaching of soils, and non-point releases from residential and commercial uses of mercury products. Non-point releases include improper outdoor disposal of thermostats, automotive switches, fluorescent lamp disposal, and mercury cell batteries (Washington State Department of Ecology and King County, 2011). The Record of Decision for OU B Marine (U.S. Environmental Protection Agency, 2000) identified three primary modes of THg transport from the BNC that have a potential to recontaminate marine sediment that have undergone remediation. These pathways include submarine groundwater discharge directly to Sinclair Inlet, discharges from BNC industrial outfalls (primarily the dry dock sump and the steam plant discharges), and discharge through BNC stormwater systems. Quantification of the loadings of filtered total mercury from sources outside of the BNC is reported to put the BNC loadings into environmental perspective.

Concentrations of Filtered Total Mercury

Generally, the mass of a source or sink is calculated by multiplying the median FTHg concentration for a source by its annual flows. In the case of the BNC dry dock discharges and the tidal flushing of stormwater drain PSNS015, the net mass is calculated to account for the recycling of Sinclair Inlet seawater into BNC structures. The mass of FTHg added from the atmosphere and sediment is calculated by multiplying a flux by the area of Sinclair Inlet.

Seawater Transported to Puget Sound

The median concentration of FTHg in the upper layer of marine water in greater Sinclair Inlet measured in this study was 0.33 ng/L (table 3). The mean value of 0.89 ng/L for 29 measurements of FTHg+PTHg by USGS was similar to the mean value of 1.09 ng/L for 17 measurements of WTHg by the ENVVEST project (table C1). The mean, median, and maximum values of WTHg measurements by the ENVVEST project that focused on storm events were higher than comparable values of FTHg+PTHg monthly measurements by USGS.

Water Discharged to Greater Sinclair Inlet

Although this report focuses on sources of FTHg in waters discharged from the BNC, FTHg was measured in effluents of wastewater treatment plants, stormwater, and streams in the Sinclair Inlet basin in 2008 and 2009 outside of the BNC as part of a companion methylmercury survey. FTHg discharged by the streams, stormwater drains, and wastewater treatment plants originate from rainfall on the landscape of Sinclair basin, leaching of THg from soils to groundwater, and use of mercury by industry, commerce, and a variety of consumer products households within the basin.

Although THg in wet precipitation samples includes a small amount of particulate THg scavenged by precipitation, wet deposition of WTHg was assumed to be filterable (FTHg). Since streamflow and stormwater originates from rainwater, the average concentration of 5 ng/L of WTHg in rainwater can be compared to FTHg concentrations in freshwater sources. If FTHg concentrations in freshwater sources are considerably less than 5 ng/L, then a portion of WTHg in the rainwater has been scavenged between the time rainwater falls on the pervious and impervious land surfaces and the water enters a creek or storm drain. Likewise, if FTHg concentrations are considerably greater than 5 ng/L, then FTHg mercury has been added to the original rainwater through leaching from soils or from anthropogenic sources. This comparison of concentrations is a simplified comparison of atmospheric deposition of THg ($M L^{-2} T^{-1}$) with basin yields of FTHg ($[M T^{-1}]/L^{-2}$).

Creek Basins

Water from four creeks (Blackjack, Anderson, Annapolis, and Gorst Creeks) within the Sinclair Inlet basin and Olney Creek (fig. 2) was sampled in May and July 2008, but the July data are qualified as estimated. Streamflow discharges in May and July differed by less than 20 percent, except that the discharge of the smallest creek sampled (Annapolis Creek) in July was one-quarter of the discharge in May. The median FTHg concentration in creek water was 0.57 ng/L and ranged between 0.39 and 0.81 ng/L (table 4). The concentrations of FTHg in creek water collected during dry weather base flow were significantly less than WTHg concentrations in rainfall indicating that the basins retained a significant portion of the atmospheric THg in their soils. The sums of FTHg+PTHg concentrations for the May and July 2009 sampling (table C1) were comparable to or slightly less than the range of THg of unfiltered water collected by the ENVVEST project during dry base flow in September 2002 (ENVironmental inVESTment Project, 2006).

Table 4. Concentrations and loadings of filtered total mercury for Sinclair Inlet, Kitsap County, Washington.

[Category: I, less than 1 gram per year; II, single digit grams per year; III, tens of grams per year; IV, hundreds of grams per year. **Abbreviations:** km^2, square kilometer; m^3/s, cubic meter per second; ng/L, nanogram per liter; g/yr, gram per year; GW, groundwater; PSNS&IMF, Puget Sound Naval Shipyard and Intermediate Maintenance Facility; DD, dry dock; FTHg, filtered total mercury; NA, not applicable; <, less than; –, not determined]

| Source | Area (km^2) | Flow (m^3/s) | Number of samples | Filtered total mercury | | | | Category |
| | | | | Concentration (ng/L) | | Loading (g/yr) | | |
				Median	Range	Median	Range	
Exchange from Puget Sound by difference	–	–	–	–	–	420	200–800	NA
Advection from Puget Sound	NA	98	40	0.20	<0.1–0.6	620	–	NA
Advection to Puget Sound	NA	100	–	0.33	–	1,040	–	NA
Direct atmospheric deposition	8.37	0.31	–	4.75	2.16–11.3	34.6	–	III
Greater Sinclair Inlet	86.3	1.82	–	–	–	48.9	–	NA
Creeks	74.0	1.51	10	0.57	0.39–0.81	27.1	24.2–34.2	III
Stormwater	12.3	0.093	3	4.25	4.02–4.62	12.3	11.6–13.4	III
Municipal effluent	NA	0.22	3	1.38	1.06–1.55	9.5	7.3–10.7	II
Bremerton navel complex sources	1.52	0.43	NA	NA	NA	146+Capture Zone GW	–	NA
Zone of direct discharge	0.10	0.009	NA	NA	NA	6.74	–	NA
Groundwater	NA	0.006	10	[1]4.06	[2]1.4–6	[1]0.81	[2]0.3–1.2	I
Stormwater	NA	0.001	0	[3]1.83	1.5–2.16	0.03	0.03–0.04	I
Steam plant	NA	0.002	4	78	15.5–143	5.9	1.2–12	II
Vicinity of Site 2	0.43	0.049	NA	NA	NA	120.2	–	NA
Groundwater	NA	0.0002	5	194	72–702	1.2	0.6–2.5	II
PSNS015 Stormwater drain	NA	0.049	NA	NA	NA	119	–	IV
Freshwater	NA	0.010	1	144	NA	46	–	NA
Tidal flushing	NA	0.039	1	[4]58.8	NA	73	–	NA
Capture zone of sumps	0.99	0.368	NA	NA	NA	18.8+GW	–	NA
PSNS&IMF stormwater	NA	0.0083	2	1.83	1.50–2.16	0.48	0.39–0.57	I
Dry Docks	NA	0.36	NA	NA	NA	18.3	–	III
DD 1-5	NA	0.16	7	[1]1.36	[2]0.63–4.16	6.9	[2]3.2–26	NA
DD 6	NA	0.20	4	[1]1.81	[2]0.97–2.38	11.4	[2]8.2–13.9	NA
Groundwater	NA	unknown	10	2.62	1.07–31	Thought to be minimal, but site 10C and 10E need further investigation		
Diffusion from marine sediments	8.37	NA	24	9.30	5.07–25.2	100s	100s	IV
Total Sinclair Inlet sources	–	–	–	–	–	230+100s	–	–

[1]Using concentrations of filtered total mercury from Wisconsin Mercury Research Laboratory.

[2]Range calculated from median by setting non-detectable values from the National Water Quality Laboratory to 0 and to 6 ng/L.

[3]Using FTHg concentrations from PSNS&IMF stormwater.

[4]Flux-weighted average during the March 31, 2010, ebb cycle.

Stormwater

Navy City stormwater drain in Gorst (fig. 4) was sampled at low tide during a heavy storm event on January 7, 2009, resulting in a specific conductance measurements of 74 µS/cm and a FTHg concentration of 4.25 ng/L. The stormwater drains near Sheridan Road (fig. 1) and near the boat ramp of the Port Orchard Marina (fig. 4) were sampled at low tide the following evening (January 8, 2009) as the trailing edge of the storm passed, resulting in values of specific conductance measurements of 20 and 75 µS/cm, respectively, and FTHg concentrations measurements of 4.62 and 4.02 ng/L, respectively. FTHg concentrations in stormwater were comparable to those in rainwater and indicate that little FTHg is scavenged during storm events similar to the event of January 7, 2009. The sum of the FTHg+PTHg concentrations in the Navy City stormwater drain (table C1) was within the range of WTHg concentrations measured between 2002 and 2005 (ENVironmental inVESTment Project, 2006).

Wastewater Treatment Plant Effluents

FTHg and PTHg were measured in May, July, and August 2008 from a composite of five grab samples collected over the 8-hour day shifts from the wastewater treatment plant effluents for the City of Bremerton and West Sound Utility District. The median FTHg concentration for the City of Bremerton Wastewater Treatment plant (table 4) was 1.38 ng/L (range: 1.06 to 1.55 ng/L). USGS THg data collected during non-storm periods in the summer of 2008 were less than comparable data collected by ENVVEST (ENVironmental inVESTment Project, 2006) mainly during storm events (table A9). The effluent from the West Sound Utility District does not discharge into the Sinclair Inlet model box; results are presented only for comparison with the City of Bremerton plant. The median FTHg concentration in effluent from the West Sound Utility District treatment plant was 2.21 ng/L (range: 2.20 to 3.52 ng/L).

Water Discharged from the Bremerton Naval Complex

FTHg discharged to Sinclair Inlet from the BNC originates from current naval operations and leaching of legacy mercury from contaminated soils by fresh and saline groundwater within the BNC. Within the terrestrial boundary of the BNC, FTHg is also added from rain falling onto the BNC landscape, seepage of seawater from OU B Marine into the dry docks, and seepage of ambient fresh groundwater entering the landward boundaries of the site.

The USGS collected groundwater samples for FTHg analyses from 12 monitoring wells (fig. 4) between January and June 2008 to evaluate dispersed and point (localized) sources of THg to Sinclair Inlet from groundwater beneath the BNC. During the first base-wide sampling event (January 28–February 1), FTHg was measured by the USGS NWQL. Insofar as FTHg concentrations above the estimated level of 6 ng/L were measured in only 3 of the 11 samples, FTHg in samples from the second base-wide sampling event in April were analyzed by the WMRL that reported FTHg concentrations as low as 0.04 ng/L. The USGS also conducted more intensive groundwater sampling over selected tidal periods during May and June 2008 at wells OUBT-722 and OUBT-715R, and at piezometers in the intertidal zone adjacent to OUBT-722 (fig. 5 inset).

Zone of Direct Discharge

Groundwater in western BNC is thought to discharge directly to Sinclair Inlet (Prych, 1997). Groundwater samples were collected from four monitoring wells in OU A by the USGS during January and April 2008, and well OUBT-715R was sampled in May 2008. Groundwater from well OUA-206 contained freshwater, whereas the four other wells contained saltwater. During the April sampling events, groundwater levels and salinities were lower than those during the January sampling event. Samples collected from OU A in January were analyzed by NWQL with the reporting level of 10 ng/L, as were the two samples collected from OUBT-715R in May. Only the groundwater collected from well OUA-241 in January contained a FTHg concentration at the reporting level. Calculating median values of data sets containing non-detectable concentrations with varying reporting levels is problematic (Helsel and Hirsch, 1992). In this report, the median value for each data category was calculated using only data from the WMRL. The median of the four detectable FTHg concentrations collected from the Zone of Direct Discharge in April analyzed by WMRL was 4.06 ng/L (table 5). The lower range of the median FTHg concentration of the entire data set is calculated by first setting non-detectable values of the NWQL to zero, then calculating the median value of the mixed data set. The upper range of the median FTHg concentration was calculated by setting non-detectable concentrations from the NWQL to the lowest estimated concentration reported by NWQL (6 ng/L), then calculating the median value for the mixed data set. The median for the mixed WMRL and NWQL data set ranged between 1.4 and 6 ng/L. Since the reporting level for the least sensitive method was 10 ng/L, the frequency of values above 10 ng/L is a robust non-parametric indicator that can be used with mixed data sets to compare FTHg concentrations across sample categories. No groundwater sample from the Zone of Direct Discharge contained FTHg in excess 10 ng/L.

Table 5.   Frequency of filtered mercury concentrations greater than 10 nanograms per liter in samples from the Bremerton naval complex, Kitsap County, Washington.

[Source of data: Huffman and others, 2010. **WMRL:** Wisconsin Mercury Research Laboratory; QA, samples measured by WMRL were collected for interlaboratory comparison of NWQL. **Mixed WMRL and NWQL data set:** NWQL, National Water Quality Laboratory; FTHg, filtered total mercury; ng/L, nanogram per liter. **Abbreviations:** OUBT, Operable Unit B Terrestrial; TW, temporary well; PSNS, Puget Sound Naval Shipyard; IMF, Intermediate Maintenance Facility; DD, dry dock; <, less than; –, no data]

Zone	WMRL			Mixed WMRL and NWQL data set			
	Number of samples	Median	Range	Number of samples	Frequency greater than 10 ng/L (percent)	Median FTHg assuming NWQL non-detectable concentrations	
						0 ng/L	6 ng/L
Zone of direct discharge							
Groundwater	4	4.06	0.65–8.33	10	0	1.4	6
Steam plant	4	78.3	15.5–143	4	100	78.3	78.3
Vicinity of Site 2							
OUBT-722	3	QA	QA	44	100	495	495
OUBT-718	1	0.8	–	2	0	0.4	4
TW-05	5	194	71.2–702	5	100	–	–
TW-01 to TW-04	40	1.3	<0.04–9.4	40	0	–	–
PSNS 015 stormwater	1	144	–	–	100	–	–
PSNS 015 tidal flushing	1	QA	QA	10	80	41	41
Capture zone of sumps							
Fresh ambient groundwater (OUNSC-380)	1	1.1	–	2	0	0.5	4
Saline groundwater from capture zone of sumps	4	2.62	1.07–6.14	8	13	0.5	6
PSNS&IMF stormwater	2	1.83	1.50–2.16	2	0	–	–
DD1–5	8	1.36	0.63–4.16	8	0	0.3	5.1
DD6	4	1.8	0.97–2.38	6	0	1.3	2.2

FTHg concentrations below 10 ng/L are consistent with the Site Inspection (1990–91), the OU A RI/FS, and the Synoptic Study (1998–2002) in which none of 71 filtered samples had unqualified detectable concentrations at reporting levels between 100 and 200 ng/L (U.S Navy, 1995a). Nearly all previous reported high concentrations of WTHg (concentrations in micrograms per liter) from OU A wells probably were not representative of aqueous phase mercury in groundwater that could migrate to Sinclair Inlet because excess solids likely were entrained in the water samples (appendix A). No localized sources of mercury were identified in OU A.

All steam plant effluent samples contained FTHg concentrations above 10 ng/L in 2008 (range: 15.5 to 143 ng/L). FTHg concentrations were correlated (correlation coefficient=0.96, p value=0.04) with specific conductance (fig. 9). In 2010, when the steam plant demineralizing process changed to a reverse osmosis system, the use of sulfuric acid and caustic soda reagents that contained trace amounts of THg was discontinued. This change will likely decrease the THg concentrations in the effluent.

Vicinity of Site 2

Data from this and several past studies described above indicate that Site 2 (fig. 5) is an area of elevated THg concentrations in a variety of water and solids for the following reasons: (1) soils to 12 m (38 ft) below the site contain high THg concentrations ranging between 6.6 and 31 mg/kg (table A1), (2) high, but questionable, concentrations of WTHg were detected in groundwater collected from wells, which since have been decommissioned, (3) high concentrations of WTHg (65.51–1,131.02 ng/L, table A10) were measured in stormwater from drain PSNS015 that passes under the site and is tidally connected to the site, and (4) high WTHg concentrations (990–6,690 ng/L, converted from concentrations in micrograms per liter from table A12) in groundwater from well LTMP-3 adjacent to the site were measured and have been verified by more sensitive analytical methods (1,190 and 3,680 ng/L, table A11).

In winter and spring 2008 events, USGS collected groundwater samples from two monitoring wells adjacent to Site 2; OUBT-722 (called LTMP-3 by the U.S. Navy) along

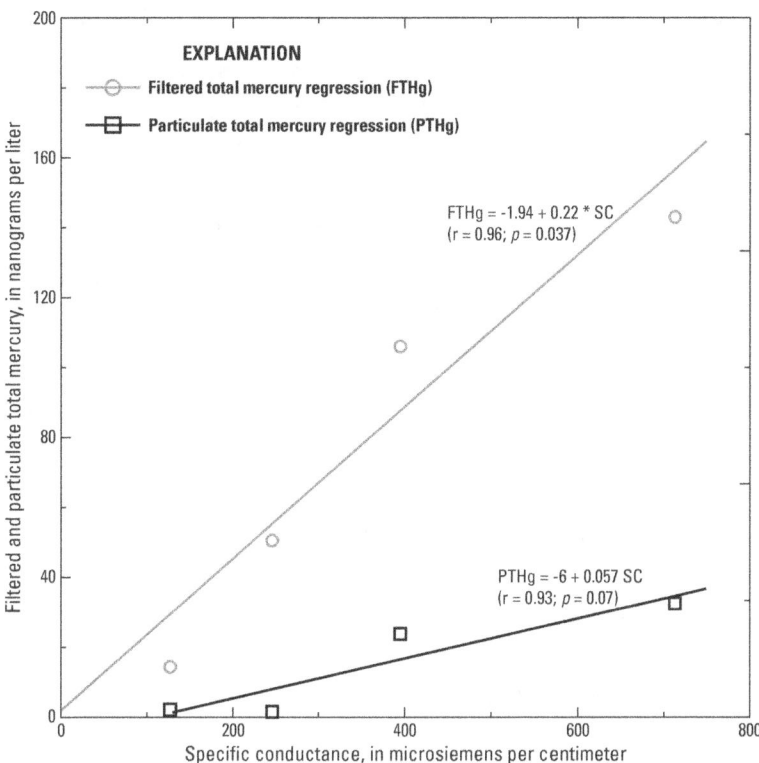

Figure 9. Filtered and particulate total mercury relative to specific conductance in the effluent from the steam plant, Bremerton naval complex, Kitsap County, Washington.

the shoreline and well OUBT-718, which is 40 m upland (fig. 4). Similar to the LTMP results, FTHg concentrations in saline oxygenated groundwater from OUBT-722 were the highest for the two USGS BNC basin-wide sampling events, 438 and 541 ng/L. In contrast, groundwater from well OUBT-718 was fresh with low concentrations of dissolved oxygen (0.5–1.0 mg/L) and an FTHg concentration of 0.8 ng/L in April, reflecting groundwater originating from the upland basin. Considering the short distance between OUBT-722 to OUBT-718, the differences in water types and concentrations of FTHg between these two wells indicated that the plume of contaminated saline water sampled at well OUBT-722 is confined to a thin strip of aquifer along the shoreline.

May 6, 2008 Tidal Study

USGS conducted intensive sampling over selected tidal periods at well OUBT-722 to better understand the potential for FTHg migration from groundwater to Sinclair Inlet. The groundwater from well OUBT-722 was sampled for about 14 hours on May 6, 2008, as water elevation in Sinclair Inlet ebbed from +3.7 m to -1.0 m (MLLW), and flooded back to +3.9 m. While water level elevations in the well and in Sinclair Inlet were monitored, FTHg and salinity were monitored hourly from sampling ports positioned at the top of the well OUBT-722 screen (+3.68 m relative to MLLW) or

just below the water elevation, at mid-depth of the screened interval (+1.45 m), and near the bottom of the screened interval (-0.6 m).

The water-level and salinity data clearly indicate that marine water saturates the soils of the site and then drains back to Sinclair Inlet in the vicinity of well OUBT-722 over a tidal cycle (fig. 10). Groundwater and Sinclair Inlet water levels were nearly identical at the 5:00 a.m. high tide. As the tide ebbed in Sinclair Inlet, the groundwater level also decreased, although at a slower rate. By about 8:00 a.m., the measurable difference between the well elevation and the Sinclair Inlet sea surface indicated that seawater was draining from the upland area. The rate of decrease in the groundwater level slowed when the tide was in full ebb, and the groundwater level stabilized at approximately +0.96 m MLLW before increasing with the flooding tide. These measurements indicate there are hydraulic characteristics in the vicinity of well OUBT-722 that allow for rapid drainage of groundwater above about +1 m, but constrain groundwater drainage below that elevation. The geology log for well OUBT-722 does show a contact between contrasting materials at an elevation of +1.38 m MLLW, but the material change is from fine sandy fill material above the contact to fine-coarse sandy fill below the contact; the hydraulic characteristics for those two types of materials likely are quite similar. Thus, manmade subsurface features such as stormwater drains or the seawall more likely are the reasons for the constrained groundwater levels below +1 m MLLW. The salinity in well OUBT-722 also decreased from approximately 24.5 to 20.5 during the later stages of the ebb tide, indicating an influx of freshwater.

FTHg concentrations in groundwater at OUBT-722 over the tidal cycle followed a pattern that was neither the same as, nor opposite from, changes in groundwater levels or salinity (fig. 10). At the three depths sampled within the well casing, FTHg concentrations were relatively low during the initial high tide (225–312 ng/L), then increased quickly during the early stages of the ebbing tide to a maximum of 802 ng/L in the sample just below water table, as saline water drained from the upland area. Throughout the ebbing tide, the highest FTHg concentration was near the top of the water column and decreased in samples collected from lower elevations within the screened interval. The timing of the increase of FTHg in saline water and the vertical distribution of the FTHg indicate that the strong source of FTHg was not marine water from Sinclair Inlet, but was groundwater drainage from near-ground surface soils in the upland area that had been flushed with seawater.

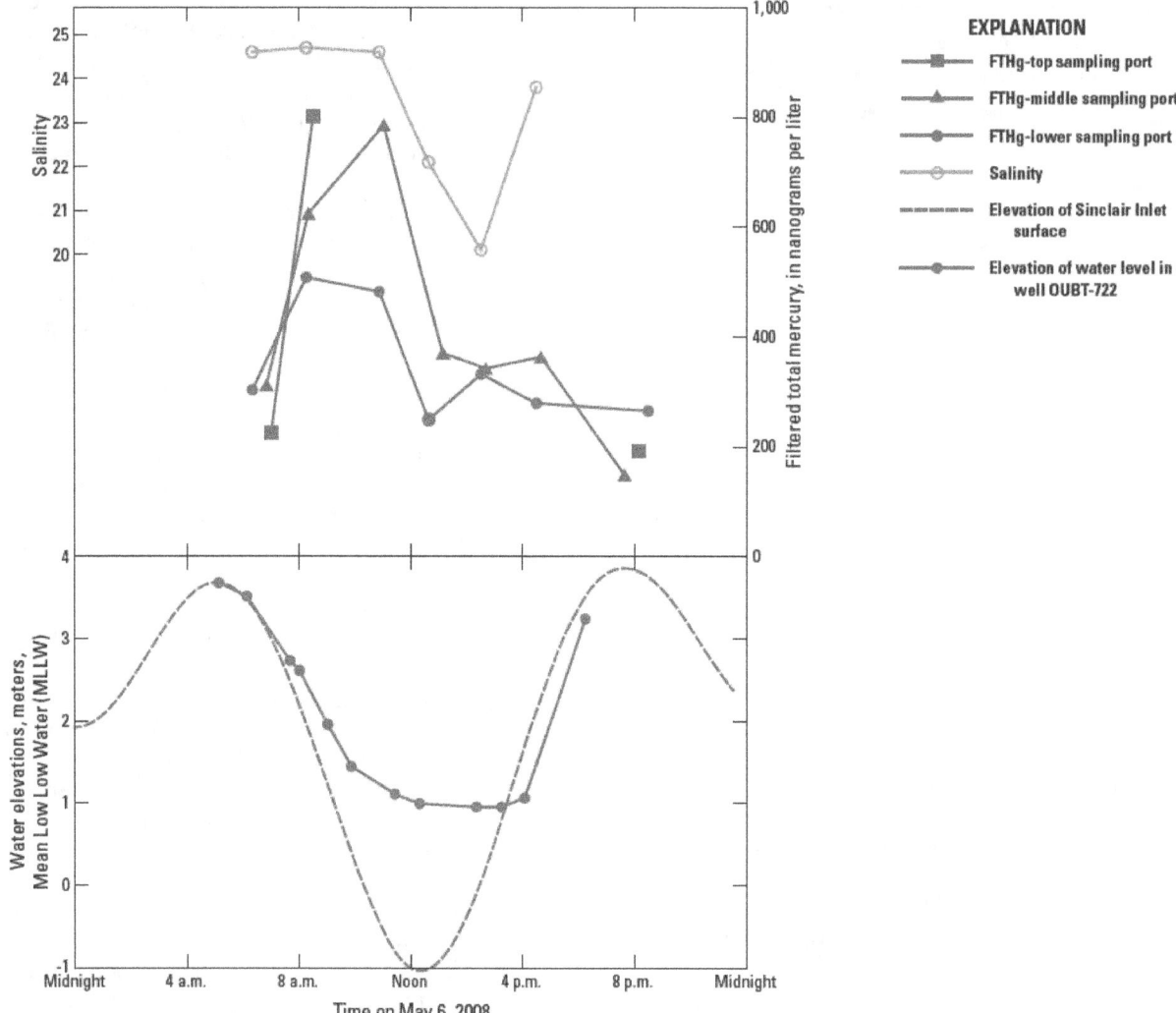

Figure 10.   Water levels, total mercury concentrations of filtered water, and salinity in well OUBT-722 during the May 6, 2008 tidal study. Salinity was obtained below the sampling port at the bottom of the screened interval, Bremerton naval complex, Kitsap County, Washington.

As ebb tide continued and water elevations in both Sinclair Inlet and the well dropped, the top sampling port was immediately above the mid-point sampling port and sampling from the top sampling port was discontinued until the following flood tide. As drainage of the upland areas continued after 11:00 a.m., maximum FTHg concentrations in groundwater decreased to 400 ng/L with the influx of freshwater, and dissolved oxygen decreased by 0.2 mg/L (not shown). During the following flood tide, groundwater salinity increased, but FTHg concentrations continued to decrease to below 250 ng/L at all elevations. The timing and vertical distribution of measured salinity and FTHg concentrations indicate that groundwater at well OUBT-722, at a given time, is a varying mixture of three end members: (1) saline groundwater draining near-ground surface contaminated upland soils that has a relatively high FTHg concentration (significant during mid-ebb tide), (2) fresh groundwater draining from upland areas that has a relatively low FTHg concentration (significant during the latter stages of the ebb tide, and (3) saline groundwater with a relatively low FTHg concentration that directly infiltrates to the well from Sinclair Inlet (significant during the flooding tide.)

One plausible mechanism to explain the May 6 measurements, is that seawater intruding into the aquifer mobilizes aquifer-sediment bound mercury (likely through geochemical processes such as complexation with chloride) from a unknown location upgradient of well OUBT-722. FTHg concentrations greater than 1,000 ng/L were measured in saline water collected from wells drilled into clastic soils in Southern Tuscany, Italy, containing mercury ores (Grassi and Netti, 2000). The geochemical modeling indicates that a high degree of complexation with seawater chloride mobilizes FHTg from cinnabar deposits that are present in the area. The mobilized FTHg passes well OUBT-722 as saline groundwater drains towards Sinclair Inlet during the ebb tide. This mechanism is consistent with the historical variability in mercury concentrations observed in filtered and unfiltered water from well OUBT-722 that was not explained by differences in turbidity of groundwater samples.

June 4–5, 2008 Tidal Study

Additional tidal-related data were collected during June 4–5, 2008 in the vicinity of well OUBT-722 to gain further insight into THg migration in groundwater. Again, groundwater in well OUBT-722 was sampled just below the fluctuating water table, and water levels in the well and in Sinclair Inlet were monitored over an approximate 30-hour period that included two ebb cycles (fig. 11). Twenty-five FTHg samples were collected with accompanying salinity data at about 1 hour increments from near the top of the fluctuating water table in well OUBT-722. Water levels in the well were measured periodically, and differences between well and Sinclair Inlet water levels were measured periodically using a manometer board.

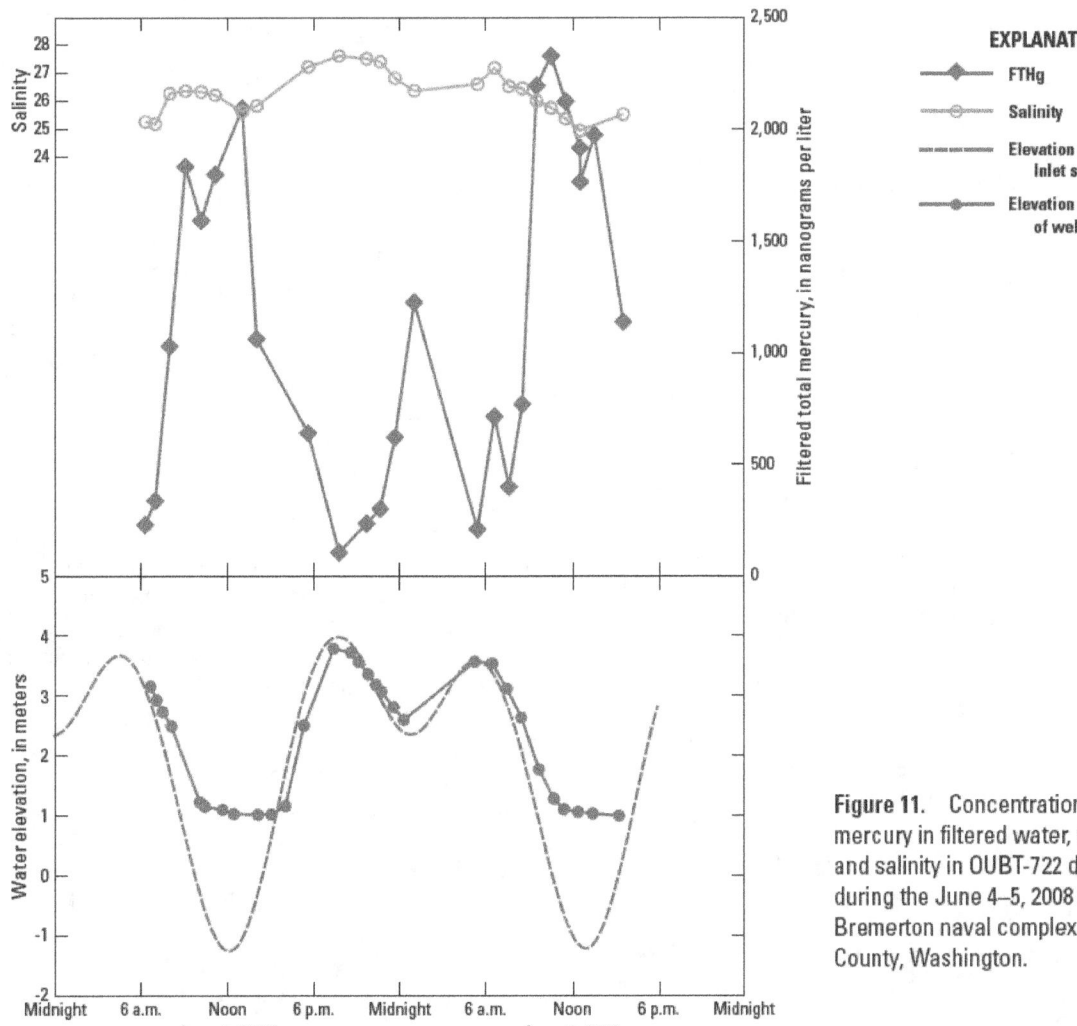

Figure 11. Concentrations of total mercury in filtered water, water levels, and salinity in OUBT-722 during the during the June 4–5, 2008 tidal study, Bremerton naval complex, Kitsap County, Washington.

The range in FTHg concentrations during the June sampling was 102 to 2,325 ng/L compared to a range of 148 to 802 ng/L during the May event. The tidal-induced changes in level, salinity, and FTHg concentration of well water measured during June 4–5 more clearly illustrate the patterns observed on May 6 (fig. 10). FTHg concentrations in groundwater from OUBT-722 spiked during the early stages of the large ebbing tides when the upland aquifer was draining, and there was even a smaller but similar response during the minor ebb in the early hours of June 5. Although the morning ebbing cycles of May 6 and June 4 were similar (+3.7 to approximately -1.2 m), the maximum FTHg concentration measured on June 4 (2,090 ng/L) was higher than the maximum concentration measured on May 6 (802 ng/L). The maximum FTHg concentration of 2,325 ng/L on June 5 occurred after a higher high tide of +4.0 m that was at a higher elevation than the May 6 higher high tide of 3.7 m. The substantial dependence of FTHg concentrations on tidal phase indicates that if maximum mercury concentrations are of interest, future sampling under the long-term monitoring program at OUBT-722 should be conducted during a specified tidal phase, preferably at the mid-point during an ebbing tide between a high-high and a low-low tide.

Prior to June 4, 2008, five piezometers were installed in the intertidal zone adjacent to well OUBT-722 in order to directly measure and estimate, if possible, the loading of FTHg in groundwater into Sinclair Inlet. The piezometers were installed parallel to the shoreline (fig. 5) to provide additional information on the direction of localized groundwater and FTHg flow. Groundwater samples were collected from most piezometers ten times during June 4–5, 2008, and FTHg was measured by WMRL; only six samples were collected from piezometer TW-T05 because it was dry during lower tidal periods. Median FTHg concentrations in water from the four western piezometers (TW-T01 through TW-T04) ranged from 0.3 to 1.7 ng/L with individual measurements ranging from less than 0.04 to 9.4 ng/L (table 6). In contrast, the FTHg concentration found in saline water collected from the eastern-most piezometer (TW-T05) nearest the seawall ranged between 71.2 and 702 ng/L, with a median concentration of 194 ng/L.

The June 4–5 data are consistent with the hypothesized mechanism developed to explain the May 6 data; seawater intrudes into the aquifer and mobilizes aquifer-sediment-bound FTHg from an unknown location upgradient of well OUBT-722, and the mobilized FTHg in saline groundwater then drains seaward past well OUBT-722 during the ebb tide. The prevalent high concentrations of FTHg at the eastern-most piezometer near the seawall strongly suggests that seawater contaminated with FTHg largely originates from the direction of the PSNS015 stormwater drain, flows along the seawall, then flows around the end of the seawall to Sinclair Inlet in the vicinity of piezometer TW-T05.

Table 6.   Summary total mercury and methylmercury concentrations in filtered water, salinity, and particulate total mercury from five piezometers installed in the intertidal zone seaward of well OUBT-722, Bremerton naval complex, Kitsap County, Washington.

[Source of data: Huffman and others, 2012. Concentration of total mercury on particles was calculated by dividing the concentration of total mercury in nanograms per liter by the concentration of total suspended solids in milligrams per liter. The units of nanograms per milligram and milligrams per kilogram are the same. **Abbreviations:** ng/L, nanogram per liter; mg/kg, milligram per kilogram; N, number of samples]

	Piezometer				
	TW-T01	TW-T02	TW-T03	TW-T04	TW-T05
Salinity					
Median	27.1	26.6	27.4	27.3	27.2
Minimum	25.9	24.9	26.6	27.1	26.8
Maximum	29.0	27.2	28.2	28.7	28.0
Number of samples	10	10	10	10	6
Total mercury, filtered (ng/L)					
Median	1.7	0.3	1.1	1.7	194
Minimum	0.6	0.2	<0.04	0.9	71.2
Maximum	3.1	9.4	2.0	2.3	702
Number of samples	10	10	10	10	6
Methylmercury, filtered (ng/L) (N =1)					
	0.165	0.09	0.08	0.28	1.5
Calculated concentration of total mercury on particles (mg/kg) (N = 2)					
Median	1.98	4.20	2.58	2.21	86.3
Minimum	1.41		1.16	1.85	75.6
Maximum	2.54		3.99	2.56	96.9
Number of samples	2	1	2	2	2

Mercury in the Stormwater Drain PSNS015

The largest drain in the BNC (PSNS015) located near well OUBT-722 (fig. 5) was sampled by USGS during one heavy storm event on January 7, 2009, at low tide resulting in the collection of non-saline water (specific conductance of 28 µS/cm). A FTHg concentration of 144 ng/L was measured and the sum of FTHg+PTHg (366 ng/L) for this event was within the range of WTHg concentrations (15 to 1,131 ng/L) for storm events reported by ENVironmental inVESTment (2006) for samples collected between 2002 and 2005 (table C1).

A key processes leading to the discharge of FTHg to Sinclair Inlet may be the tidally-driven pumping of seawater into the PSNS015 stormwater drainage system during high tides, followed by leakage of seawater into THg-contaminated soils of the aquifer; geochemical extraction of soil THg by seawater; and subsequent drainage back into the stormwater drain or westward transport to Sinclair Inlet through the aquifer behind the 75 m portion of the seawall and past well OUBT-722 and TW-05 where the quay is absent. To test this hypothesis, USGS sampled PSNS015 from the manhole PSNS015-2253 (fig. 5) near the seawall during the ebbing cycle from the higher high tide of +3.83 m at 6:10 a.m. of March 31, 2010, after only 0.13 cm of rain had fallen in the previous 24 hours and there was only a slight chance of rain. Because the plug flow of Sinclair Inlet seawater into the stormwater drain during the previous flood cycle would transport the first water passing by the manhole as it flowed out of the stormwater drain, we focused sampling during the last 3 hours of the ebbing tide when groundwater most likely would drain back into the stormwater drain pipe.

The sampling port was positioned within the manhole vault at the top of the 1.22-m pipe of the stormwater drain. Salinity decreased from 14.8 at 9:50 a.m. to approximately 6.1 at 11:15 a.m. as water level in the manhole vault decreased from +1.63 m above the top of the pipe to +0.42 m (fig. 12). Salinity remained stable at about 6 for the next 45 minutes as the water level dropped to the top of the pipe and below the elevation of the sampling port at about noon. The sampling port was lowered to the middle of the pipe, where salinity was 27.88. Light rain started falling at 11:50 a.m. and stopped at 12:10 p.m., but salinity did not decrease appreciably at the center of the pipe (noon–12:10 p.m.) during the first 10 minutes. As the water level continued to drop within the stormwater drain over the next 40 minutes, salinity decreased to about 21. During the subsequent flooding tide starting at 1:00 p.m., salinity increased to 28.5. The vertical structure of salinity in the manhole vault clearly indicates that the water in the stormwater drain was stratified and the stormwater drain system is acting as a confined estuary.

Between 10:16 a.m. (flow of 0.17 m³/s) and 11:52 a.m. (flow of 0.15 m³/s), FTHg concentration measured at the top of the pipe decreased from 75.4 ng/L to 49.7 ng/L (fig. 13). Water initially sampled from the center of the pipe contained a FTHg concentration of 39.8 ng/L (salinity of 27.88) and decreased to 27.9 ng/L (salinity of 21.85). After the water had drained from the manhole vault and was confined to the pipe, flow rapidly slowed during the final hour of the ebbing tidal cycle. Between 12:40 p.m. and 1:00 p.m. as water levels reached their lowest elevations during slack tide and flow ceased, FTHg increased to 89.4 ng/L. FTHg decreased to undetectable concentrations (10 ng/L) at the center of pipe as new Sinclair Inlet water flowed into the pipe during the subsequent flooding tidal cycle. Because of changes in flow over time, the FTHg concentration of 20.2 ng/L measured in the time-averaged composite was much less than the flow-weighted concentration of 58.8 ng/L derived from the individual hourly concentrations and flow measurements.

As with the groundwater near OUBT-722, three water masses are inferred in the drain PSNS015: (1) saline water containing high concentrations of FTHg (observed at the beginning of the study); (2) high salinity, low-FTHg water from Sinclair Inlet (observed at the end of the study); and (3) brackish water containing high concentrations of FTHg (after the water drops below the top of the pipe). Unlike the groundwater OUBT-722, the brackish water in the stormwater drain contained high concentrations of FTHg. The changing position of the sampling, changing salinity and changing flow during this single 3 hour experiment do not allow for a comprehensive understanding of the physical and geochemical processes controlling the observed FTHg concentrations. However, it is clear that FTHg concentrations of saline water in the stormwater drain were higher than concentrations in any groundwater sampled outside of the Site 2 area.

Capture Zone of Sumps

Most of the groundwater in OU NSC and the PSNS&IMF is captured by the zone of depression of the drainage systems of the six dry docks (Prych, 1997). As with groundwater from OU A, well water levels measured in April 2008 were lower than those measured in January 2008. Groundwater collected from the deepest well (OUNSC-380) in April was the freshest groundwater sampled, and thus, was assumed to represent the background groundwater FTHg concentration (1.1 ng/L) for the BNC. The median FTHg concentration in saline groundwater measured by WMRL in April was 2.62 ng/L and concentrations ranged between 1.07 ng/L (OUBT-406R in April) and 31 ng/L (OUBT-709 in January). Median concentrations for the eight samples from Capture Zone of the Sumps ranged between 0.5 and 6 ng/L depending on whether non-detectable concentrations were set to 0 or 6 ng/L.

Only one well in the PSNS&INF (OUBT-724, also called LTMP-5) was sampled by both USGS and the U.S. Navy as part of the LTMP. Well OUBT-724 is located in fill material and the surrounding area may be at the edge of the zone of depression of the dry docks. WTHg concentrations of 41.6 and 27.5 ng/L were measured in groundwater from LTMP-5 by the LTMP in October 2008 and April 2009 using sensitive methods, respectively (table A11). These concentrations were considerably higher than those measured by USGS in 2008 (table C1). This difference between USGS data and LTMP data may be a result of sampling techniques that resulted in lower TSS concentrations by USGS or by natural variations caused by varying water elevations. A high concentration of WTHg (5,240 ng/L after conversion of units from table A12) in groundwater collected from OUBT-724 was measured by the LTMP on October 18, 2005 after an extreme high tidal stage. Therefore, groundwater containing elevated FTHg may be discharging from this fill area.

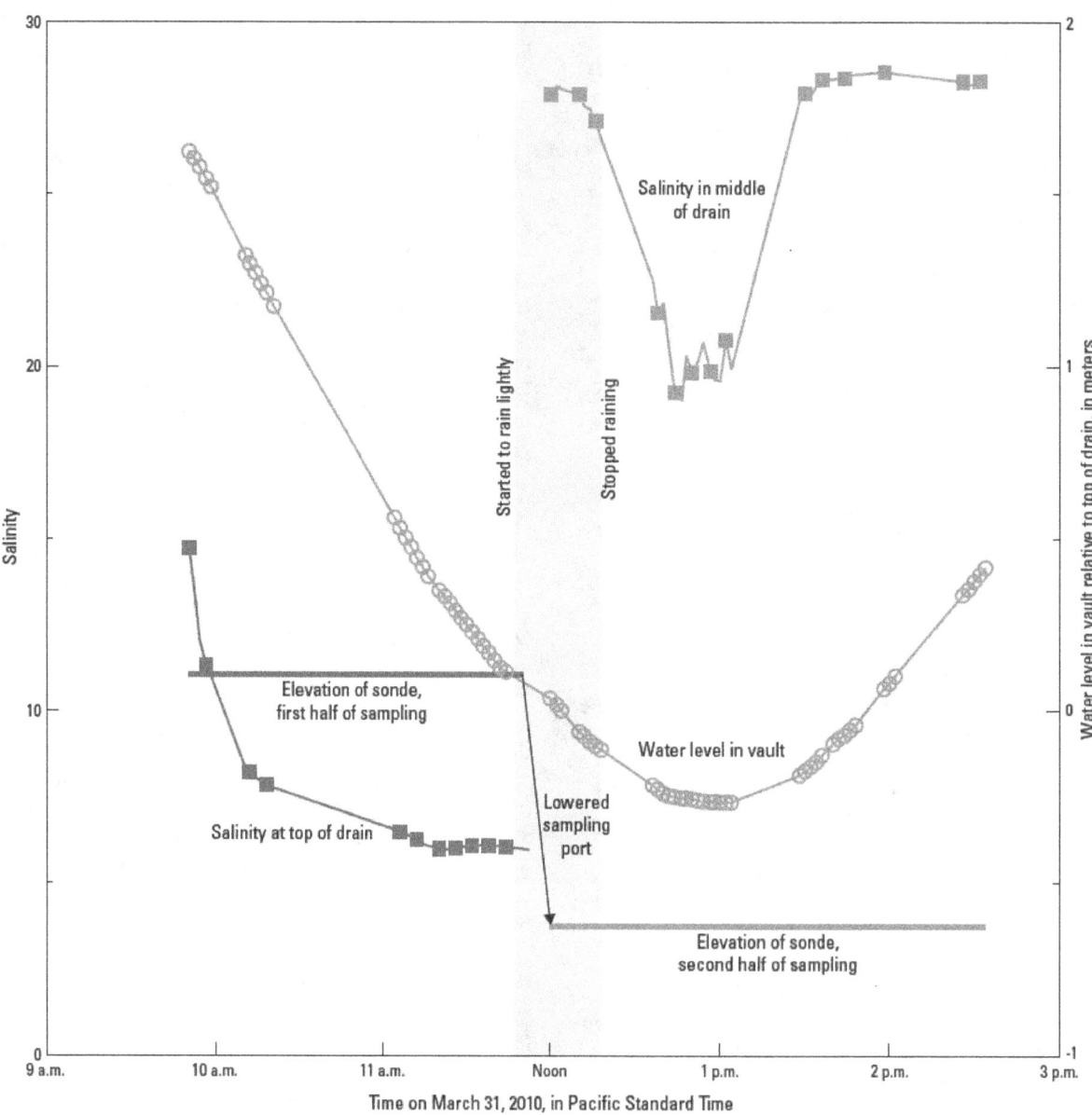

Figure 12.   Water levels and salinity in stormwater drain PSNS015 during the latter part of the morning ebbing tide on March 31, 2010, Kitsap County, Washington.

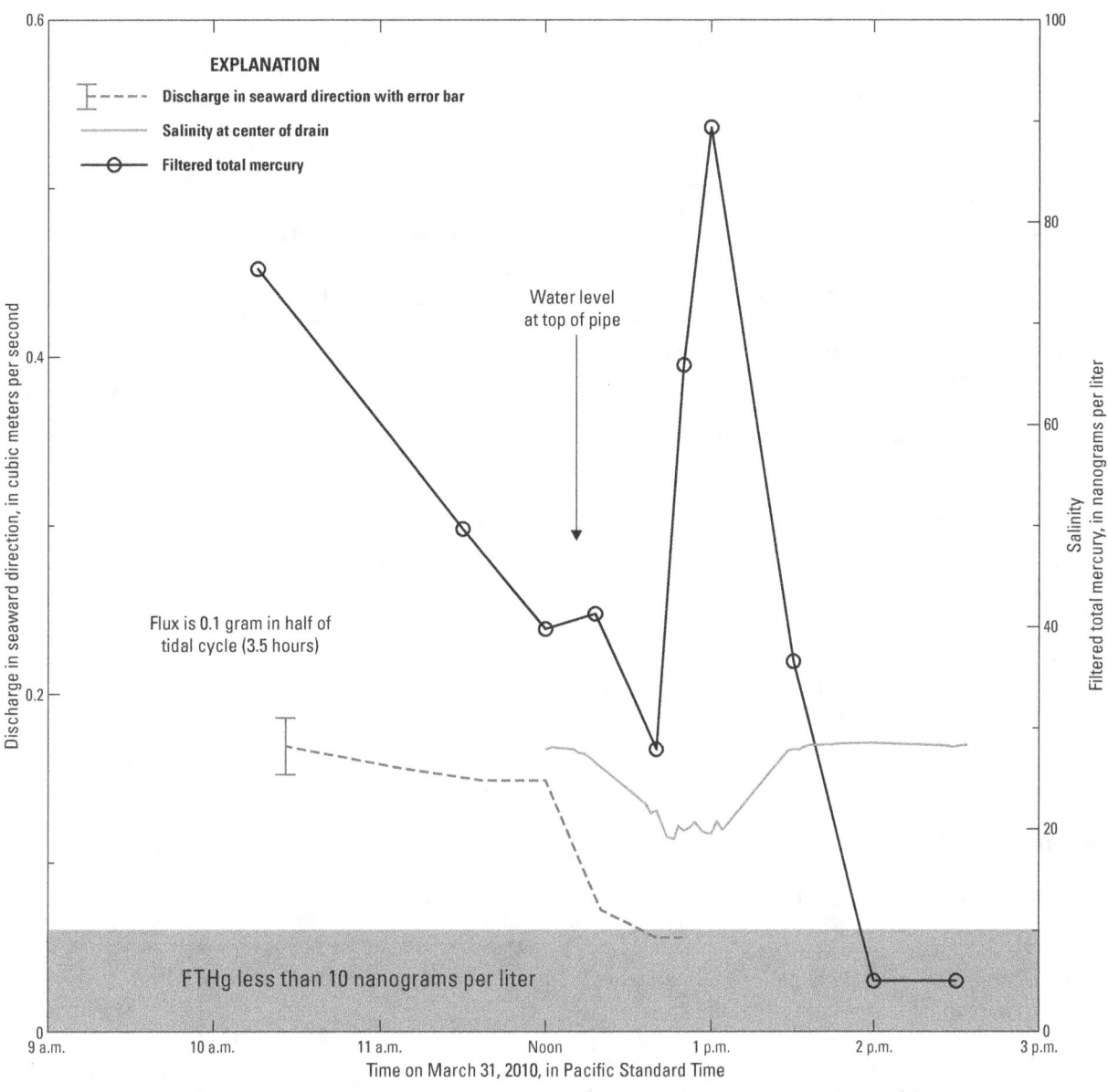

Figure 13. Flow, filtered total mercury, and salinity in the PSNS015 stormwater drain on the Bremerton naval complex during the morning ebbing tidal cycle of March 31, 2010, Kitsap County, Washington. Gray area represents the reporting level for filtered total mercury.

In the PSNS&IMF, FTHg was measured in stormwater collected from drains on both sides of a dry dock at low tide during a heavy storm event on January 07, 2009, when specific conductance of stormwater in stormwater drains PSNS 124.1 and PSNS124 was 77 and 29 µS/cm, respectively. The median FTHg concentration was 1.83 ng/L (table 5) and ranged from 1.50 to 2.16 ng/L. USGS results were within the range measured by ENVironmental inVESTment (2006) between 2004 and 2005 (table C1).

USGS sampled sump water from two dry dock relief drainage systems during 2007 and 2008. FTHg was not detected in any of the six dry dock samples collected between December 2007 and February 2008 that were analyzed by the NWQL (reporting level of 10 ng/L), and no FTHg concentration from a dry dock sample was greater than 10 ng/L (table 5). The median FTHg concentration of Dry Docks 1-5 discharge water analyzed by the WMRL was 1.36 ng/L and FTHg concentrations ranged between 0.63 and 4.16 ng/L. The median FTHg concentration of the mixed NWQL and WMRL data set ranged from 0.3 to 5.1 ng/L, depending on whether the non-detectable concentrations from the NWQL were assumed to be 0 or 6 ng/L. FTHg concentrations in Dry Dock 6 discharge water analyzed by the WMRL ranged from 0.97 to 2.38 ng/L. The median FTHg concentrations of the mixed NWQL and WMRL data set ranged from 1.3 to 2.2 ng/L, depending on whether the non-detectable concentration from the NWQL was assumed to be 0 or 6 ng/L.

Porewater of Sinclair Inlet Sediment

Four times between August 2008 and August 2009, FTHg concentrations in sediment porewater in the top 2 cm of sediment from three BNC stations and three greater Sinclair Inlet stations were measured as part of the Methylation and Bioaccumulation Project (data reported in Huffman and others, 2012). FTHg concentrations in porewaters ranged between 2.23 and 59.8 ng/L and were not correlated with THg concentrations of sediment (STHg). The median porewater FTHg concentrations for SI-IN and SI-PO stations were 25.20 and 17.54 ng/L (table 7), respectively. The lowest FTHg porewater concentrations generally were measured in February 2009, when cooler temperature and depleted nutrients limited biological activity in the water column and the sediment.

Synthesis of Loadings of Filtered Total Mercury to Sinclair Inlet

Exchange between Sinclair Inlet and Puget Sound

The difference in the transport of FTHg into and out of Sinclair Inlet (table 4) provides an estimate of the atmospheric, terrestrial and sedimentary sources added to Sinclair Inlet:

Table 7.   Concentrations of filtered total mercury in porewater of sediment and fluxes of filtered total mercury from sediment in Sinclair Inlet, Kitsap County, Washington between August 2008 and August 2009.

[Source of data: Huffman and others, 2010. **Field identifier:** SI, Sinclair Inlet; BNC, Bremerton naval complex; OUT, Outer; IN, Inner; PO, Port Orchard. Numbers after BNC stations refer to the grid cell shown in figure 5. **Abbreviations:** cm, centimeter; ng/L, nanogram per liter; ng m^{-2} day^{-1}, nanogram per square meter per day]

Field identifier	Number of measurements	Median	Range
Filtered total mercury porewater concentrations (top 2 cm), ng/L			
SI-OUT	4	9.30	7.87–10.9
SI-IN	4	25.20	2.95–59.8
SI-PO	4	17.54	4.56–35.9
BNC-39	5	5.18	2.86–16.9
BNC-52	1	6.21	
BNC-60	3	4.07	2.72–41.7
BNC-71	4	11.60	2.23–22.7
Flux[1] of filtered total mercury, ng m^{-2} day^{-1}			
SI-OUT	10	38	ND–94
SI-IN	9	269	E14–1,462
SI-PO	9	161	22–492
BNC-39	10	22	(-76)–49
BNC-52	1	0	
BNC-60	8	92	73–333
BNC-71	10	46	17–100
Median		46	

[1] The median flux the first measurements of fluxes from cores collected in August 2008, February 2009, June 2009 and August 2009. In August 2008, the flux from a single core during 2 days after collection was used. For 2009, the median of the first measurement from triplicate cores was used. The median values for the four sampling periods were used to calculate the range.

$$\left[C_{SI} \times \left(Q_{SW} + Q_{FW} \right) \right] - \left[C_{PS} \times Q_{SW} \right] = \quad (6)$$
atmospheric, terrestrial and sedimentary sources,

where

C_{SI} and C_{PS} are concentration in Sinclair Inlet and Puget Sound, respectively, and

Q_{SW} and Q_{FW} are consistent with the values of the flow of saltwater and freshwater derived in equation 5.

Transport of Puget Sound seawater into the bottom layer of Sinclair Inlet (98 m^3/s) containing an FTHg concentration of 0.2 ng/L adds 620 grams of FTHg per year of FTHg to Sinclair Inlet (table 4). Sinclair Inlet water in the upper layer of Sinclair Inlet with a median FTHg concentration of 0.33 ng/L that is leaving Sinclair Inlet (100 m^3/s) transports 1,040 grams of FTHg per year out of Sinclair Inlet to Puget Sound.

The increase of 0.13 ng/L in the FTHg concentration in Sinclair Inlet water above Puget Sound seawater indicates that an additional FTHg loading of approximately 420 grams of FTHg per year is added to Sinclair Inlet by atmospheric, terrestrial, and sedimentary sources (fig. 14). Terrestrial sources within the Sinclair Inlet basin include creeks and stormwater, municipal effluents, and BNC sources (BNC groundwater, stormwater associated with freshwater and

tidal flushing, and industrial effluents). Insofar as the average annual transport is based on limited ADCP data record and an incomplete knowledge of the factors controlling transport, both the transport of Puget Sound seawater to Sinclair Inlet and Sinclair Inlet sources may vary by a factor of 2. Since the transport of Puget Sound seawater only is known within about a factor of 2, the estimated loading from Sinclair Inlet sources ranges from about 200 to 800 grams of FTHg per year.

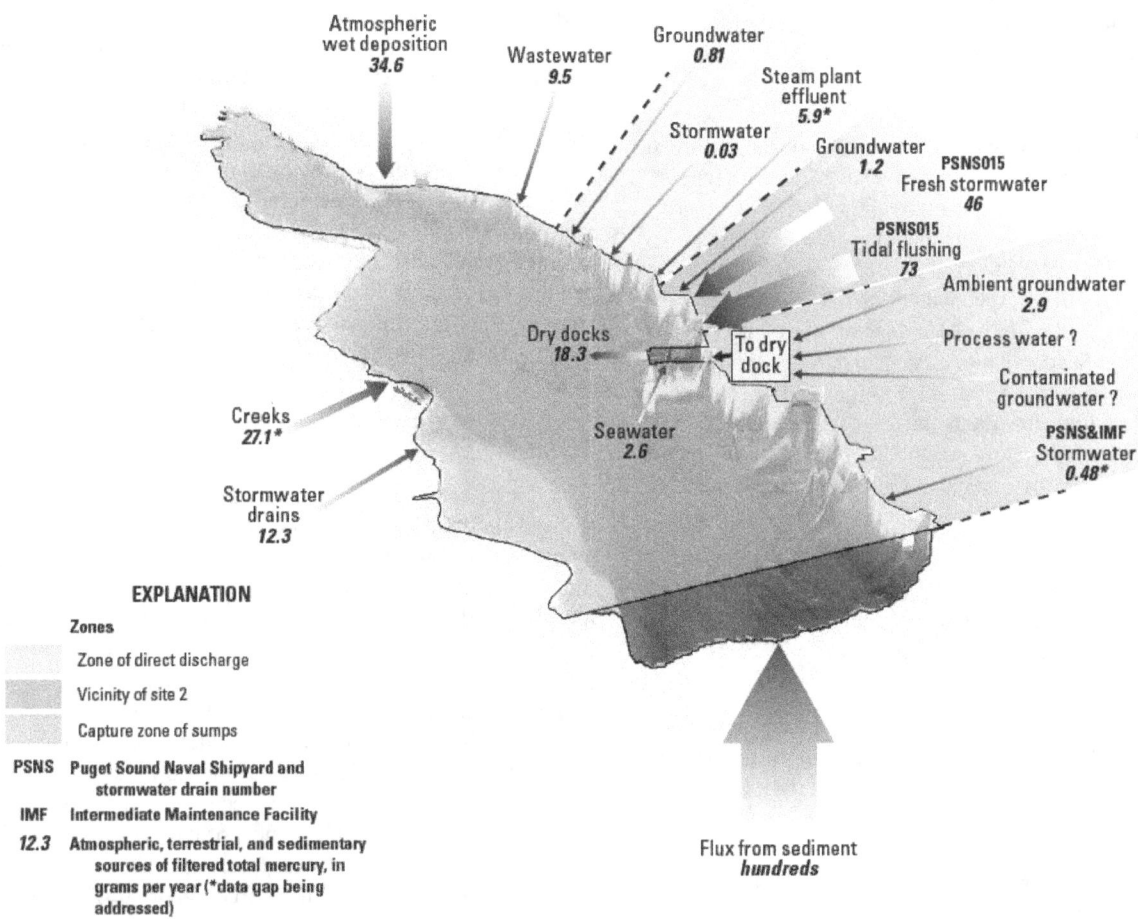

Figure 14. Loadings of filtered total mecury from freshwater sources, from seawater recycled in and out of Bremerton naval complex, and from net advective transfers between Sinclair Inlet and Puget Sound, Washington.

Direct Atmospheric Deposition

THg deposition at Manchester, Washington, located just outside of Sinclair Inlet averaged 3.23 µg m^{-2} yr^{-1} over the 18 sampling events with a minimum of 0.51 µg m^{-2} yr^{-1} and a maximum deposition of 9.38 µg m^{-2} yr^{-1} (Brandenberger and others, 2010). THg deposition during three 2-week sampling events when no rain fell averaged 0.65±0.11 µg m^{-2} yr^{-1} and provides an estimate of dry deposition. Because Brandenberger and others (2010) demonstrated a linear relationship between the cumulative precipitation per event and the THg flux per event for the Manchester site, these data were examined to determine how representative precipitation conditions were during the 18 sampling events. The 18 sampling events represented 264 days, 44 cm of rainfall, and 2.96 µg m^{-2} of THg deposition. The 44 cm of cumulative rainfall during the events is well below the annual averages of 97, 119, and 130 cm for 2008, 2009, and the long-term average for Port Orchard, respectively. Simply scaling up the measured 2.96 µg m^{-2} to annual rainfall yields annual THg depositions of 5.02, 6.21, and 6.74 µg m^{-2}, respectively. The 6.21 µg m^{-2} for 2009 compares well with the THg deposition of 7 µg m^{-2} measured at the Seattle National Mercury Deposition Site (National Atmospheric Deposition Program, 2010). However, this method could overestimate the impact of dry fall THg deposition. For an alternative method, the slope of the regression of cumulative precipitation per event and the THg flux per event for the Manchester data is used:

$$\text{Annual THG wet deposition (µg m}^{-2}) = 0.04\,\text{µg}/\text{m}^2 \quad (7)$$
$$+0.035\,(\text{µg m}^{-2}\text{per cm of rain})\times \text{annual rainfall (cm)}$$
$$(R^2 = 0.59; \ p \text{ value} < 0.001).$$

The intercept value, which represents the dry deposition, is not significant and is considerably less than the measured value of 0.65 µg m^{-2} during periods of no rainfall. Using equation 7 with the average rainfall of 118 cm for water years 2002–05 results in a long-term yield of THg wet deposition of 4.14 µg m^{-2} yr^{-1}. The loading of FTHg from atmospheric deposition (L$_{ATM}$) is calculated by:

$$L_{ATM} = \text{Annual THg wet deposition} \times A_{SI} \quad (8)$$

Using the average rainfall, equation 8 yields a loading of 34.6 grams of FTHg per year for water years 2002–05.

Loads Discharged to Greater Sinclair Inlet

For streams and stormwater drain basins, the annual loading of FTHg (L$_i$) from was calculated by multiplying median FTHg concentrations ($\hat{C}_i$) by annual water transport to Sinclair Inlet ($\overline{Q}_i$).

$$L_i = \hat{C}_i * \overline{Q}_i \quad (9)$$

If the loadings of FTHg as creek discharge is less than the amount of FTHg falling on a watershed as rain, either the water was not being delivered to Sinclair Inlet or the THg is sequestered in the watershed. The relative importance of the delivery of water and the sequestering of FTHg on the loading of FTHg from basins was assessed by comparing water yields (table 2) and FTHg concentrations in the water source.

Creeks

FTHg concentrations from streams were measured as ancillary data as part of a methylmercury survey and not specifically for mass loading purposes. Although the rate of MHg production is thought to be highest under summertime conditions, FTHg concentrations during summer may not represent average annual concentrations. In May and July 2008, there were no apparent differences between the four sampled streams and the median FTHg concentration was 0.57 ng/L. This FTHg concentration is low compared to THg concentrations in rain fall and indicates that most of the THg falling on the landscape was retained by the creek watersheds. Using an average annual transport of 47.6 × 10^6 m^3 (calculated from the 1.51 m^3/s annual average flow; Skahill and LaHatte, 2007), a loading from creeks was calculated to be 27.1 grams of FTHg per year.

Stormwater

Using a median FTHg concentration of 4.25 ng/L for stormwater drain basins outside of the BNC (table 4) and an annual water transport of 2.9 × 10^6 m^3 (from an average flow of 0.093 m^3/s; Skahill and LaHatte, 2007), stormwater basins are estimated to discharge 12.3 grams of FTHg per year to Sinclair Inlet using equation 9. FTHg in three stormwater samples collected during winter storm conditions were about 4 ng/L, indicating little retention of FTHg in stormwater during the short time that stormwater flows across the landscape to Sinclair Inlet. In contrast, the water yield from stormwater basins was only 16 percent of the water yield of atmospheric deposition. Thus, most of the water that fell on stormwater basins was not transported to Sinclair Inlet as surface water, but the THg in the fraction of rainwater that was transported to Sinclair Inlet was not removed.

The estimated atmospheric deposition of FTHg to the stream and stormwater basins (86.3 km^2) discharging to greater Sinclair Inlet is 360 grams of FTHg per year compared to the loading of 39.4 grams of FTHg per year to Sinclair Inlet from these basins. Some of the remaining 320 grams of THg per year flows to Sinclair Inlet in groundwater, but the majority of the remaining THg is probably sequestered by particles and could eventually be transported to Sinclair Inlet as PTHg. It is likely that the calculated FTHg loading from stream basins was biased low because of the lack of wintertime wet period measurements of FTHg.

Municipal Effluent

Similar to the stream data, the FTHg concentrations in effluent from the City of Bremerton wastewater treatment plant were only measured during the dry season and were less than 2 ng/L. Since the effluent concentration is less than the 5 ng/L measured in atmospheric deposition, geochemical processes within the water supply basin and engineered processes occurring in water supply and waste treatment systems must remove more FTHg than is added from anthropogenic sources. Multiplying summertime FTHg concentrations by annual total discharge yields an estimated loading of 9.5 grams of FTHg per year for the City of Bremerton wastewater treatment plant according to equation 9.

Loadings from Bremerton Naval Complex

Although concentrations of FTHg in groundwater in the capture zone were measured, most of the groundwater on the BNC is confined by a seawall and is captured by the dry dock relief drainage systems and no direct groundwater discharge of FTHg flux is assumed. Only FTHg loadings from groundwater discharge from the zone of direct discharge and from the Vicinity of Site 2 are estimated.

Zone of Direct Discharge

Groundwater not constrained by the seawall and discharging from most of the shoreline of OU A and western OU B Terrestrial (0.006 m^3/s in table 2) had a median FTHg concentration of 4.06 ng/L. This groundwater contributed an annual loading of 0.81 grams of FTHg to Sinclair Inlet according to equation 9 with an estimated range between 0.3 and 1.2 grams of FTHg per year (table 4). The small amount of stormwater discharged from this BNC area contributes 0.03 g/yr, assuming the two PSNS&IMF stormwater samples are representative of this area of the BNC.

The steam plant is located in western BNC upland of the shoreline directly discharging groundwater. In 2008, when the plant contained an ion-exchange system, the steam plant discharged an average flow of 0.0024 m^3/s. FTHg concentrations ranged between 15.5 and 143 ng/L (table 4). Because of the low discharge rate, the FTHg loading was limited to 5.9 grams of FTHg per year according to equation 9 and likely is now reduced by the conversion to a reverse osmosis water-treatment process in 2010.

Vicinity of Site 2

The sources of FTHg from the Vicinity of Site 2 include groundwater discharge and sources discharged through the PSNS015 stormwater drain. FTHg discharged from the PSNS015 stormwater drain includes sources from tidal flushing of seawater into and out of the aquifer and freshwater sources.

Groundwater

The annual loading of FTHg in saline groundwater that flows around the westward edge of the seawall and past the TW-T05 piezometer was estimated using hydraulic and chemical data. Although there are probably density-dependent interactions between seawater and freshwater along the shoreline in this area, the loading of interest is related to only the seawater that moves inland during rising tides (either through the subtidal storm drain system or directly through beach sediments) and the subsequent draining of that seawater—and associated FTHg—back to Sinclair Inlet during ebbing tides. Because the seawater essentially is a constant density, Darcy's law was used to estimate the groundwater discharge:

$$Q_{gw2b} = -K_h I A_{gw2},\qquad(10)$$

where
  Q_{gw2b} is the transport of groundwater to Sinclair Inlet from the Vicinity of Site 2 for the base case,
  K_h is the saturated hydraulic conductivity of the shallow subsurface sediments,
  I is the hydraulic gradient between the edge of seawall and Sinclair Inlet, and
  A_{gw2} is the area of groundwater flow perpendicular to the shoreline in the Vicinity of Site 2 that has elevated FTHg concentrations.

Reasonable values for the Darcy's law parameters were used to generate the most likely value for Q_{gw2b}, and those values were adjusted to generate plausible minimum and maximum values of flow. For the base case, K_h was assumed to be constant throughout the vicinity of OUBT-722 with a value of 0.0134 cm/s (38 ft/d), the "expected" value for sediments in the area (Site 10W in table 5-2, U.S. Navy, 2002). The area A of groundwater flow (A_{gw2}) that has elevated FTHg concentrations represented by those measured in TW-T05 was assumed to be 21 m^2 (7 m wide and 3 m thick; fig. 5). Estimating a reasonable value for the hydraulic gradient I was more involved.

Assuming that there is substantial saline groundwater discharge to the Inlet during only about half of each day, an average daily value for I was estimated. During June 4–5, 2008, the periods of significant discharge started during the ebb that followed the 5:00 a.m. high tide and ended about 10 hours later, midway into the flood that followed the subsequent low tide (fig. 11). It is likely that there also were a few hours of less substantial discharge to the Sinclair Inlet during the smaller ebb and flood cycle to and from lower-high tide around midnight. The preliminary average value of 0.14 meter/meter (m/m) for I was estimated by averaging the hourly seaward gradients, which were calculated by dividing the differences in water levels in well OUBT-722 and the marine water surface (0.1–3 m) by the distances between the well and the marine water surface (1–16 m). This preliminary value was then reduced by half (down to 0.07 m/m) to account for 12 hours of each day when there is no appreciable groundwater discharge to the Sinclair Inlet.

The resulting estimated groundwater flux Q_{gw2b} was approximately 0.0002 m³/s (17,000 L/d). The annual loading of FTHg in saline groundwater that flows around the westward edge of the seawall and past the TW-T05 site was estimated by multiplying the annual groundwater flow (6.2×103 m³/yr) by the median FTHg concentration, measured in TW-T05 (194 ng/L) according to equation 9, to get a loading of 1.2 grams of FTHg per year. A plausible minimum value for the FTHg loading of 0.6 g/yr could be calculated by assuming the K_h value was less by a factor of two. A plausible maximum value for the FTHg loading of 2.5 grams of FTHg per year could be calculated by assuming the Kh value was equal to the highest value reported for fill material at the site (tables 3–13, U.S. Navy, 2002).

Discharge through PSNS015 Stormwater Drain

In contrast to other BNC sources, scaling up single measurements of freshwater, and saline and water releases to annual releases indicates that discharge from the PSNS015 stormwater drain system accounts for the largest release of FTHg from the BNC, and possibly the largest load of FTHg to Sinclair Inlet. At a manhole (PSNS015 A42) slightly upland from Site 2 (fig. 5), a single sample of low-conductivity stormwater collected by USGS from the PSNS015 system contained a FTHg concentration of 144 ng/L, which was consistent with the measurements of the ENVVEST project. Applying an average annual flow of 0.010 m³/s and one FTHg concentration, yields a highly uncertain FTHg loading of 46 grams of FTHg per year from fresh stormwater discharged through the PSNS015 stormwater system (table 4). The possible sources of this freshwater FTHg could include (1) side drains discharging into the vault that service parts of Site 2, (2) unknown THg sources upgradient of Site 2, or (3) residual effects of tidal flushing of the storm drain through Site 2, as discussed in the following paragraphs.

It is likely that high concentrations of FTHg in fresh and saline water in the PSNS015 stormwater drain system originate from interactions with contaminated subsurface soils at the site rather than from above ground sources. The data collected as of 2010 indicate that geochemical processes are releasing THg into both saline and fresh waters entering the PSNS015 stormwater drain system.

The size, complexity, and construction of the PSNS015 stormwater drain system, in which a 1.3-m pipe passes through the seawall at an elevation lower than MLLW, facilitates the extraction of soil THg by tidal flushing of seawater. Seawater leaking from the complex system of vertical vaults and horizontal drains near ground level during high tides, flows through the contaminated soil (fig. 15A) and extracts THg because of its chemical affinity with the chloride

ion in seawater. Some of this high FTHg saline groundwater then flows east along the seawall resulting in groundwater FTHg concentrations as high as 2,090 ng/L in upland well OUBT-722 (fig. 11) and as high as 702 ng/L in an intertidal piezometer TW-T05 (table 6). Some of the contaminated seawater flows back into the stormwater drain system and is discharged to Sinclair Inlet through the stormwater pipe independent of storm events (fig. 15B).

What essentially is a subterranean estuary is maintained in the 1.3-m storm drain pipe where brackish water sits on top of the more dense seawater flowing in from Sinclair Inlet (fig. 16). The brackish water is a mix of stormwater and contaminated groundwater that seeps back into the presumably leaky pipe. Immediately seaward of where the pipe passes through the seawall, there is 90 degree downward bend in the pipe that connects to a vertical riser. This vertical pipe then connects to a horizontal pipe that extends some distance from the shoreline. This pipe configuration likely contributes to the stability of the subterranean estuary. During the March 31, 2010, tidal sampling when the low tidal elevation in Sinclair Inlet was -0.31 m, the water level in the stormwater drain pipe was only 0.3 m below the top of the pipe. During the flooding tide, marine water from Sinclair Inlet with lower FTHg concentrations entered the bottom of the stormwater drain through the vertical riser and pushed the contaminated water both up the vault and back up the stormwater drain pipe (fig. 16A). During the normal ebbing tides on March 31, 2010, some FTHg in the brackish water leaked from this subterranean estuary to Sinclair Inlet through dispersion in the pipe, but most FTHg likely was retained in the pipe (fig. 16B).

On March 31, 2010, during 3.5 hours of one ebbing tide period when little rain fell, 1,770 m³ of saline water with a flow-weighted average FTHg concentration of 58 ng/L discharged through the seawall in the PSNS015 storm drain to Sinclair Inlet. The calculated loading of FTHg during this partial ebbing tide contributed a loading (L_E) of 0.1 grams of FTHg to Sinclair Inlet. Scaling this loading to a year yields a highly uncertain estimated loading of approximately 73 grams of FTHg per year from tidal flushing to Sinclair Inlet through the stormwater drain according to:

$$L_{TF} = N_E \times L_E, \qquad (11)$$

where

L_{TF} is the loading from tidal flushing and N_E is the number of ebb tidal cycles in a year (730).

The elevation of the bottom of the pipe at the seawall and the angle of the pipe just seaward of the stormwater drain could lead to episodic releases of FTHg. During the ebbing tides when fresh stormwater runoff fills the vaults, the hydraulic head of fresh stormwater might be sufficient to push the contaminated seawater out of the storm drains (fig. 16C).

A.

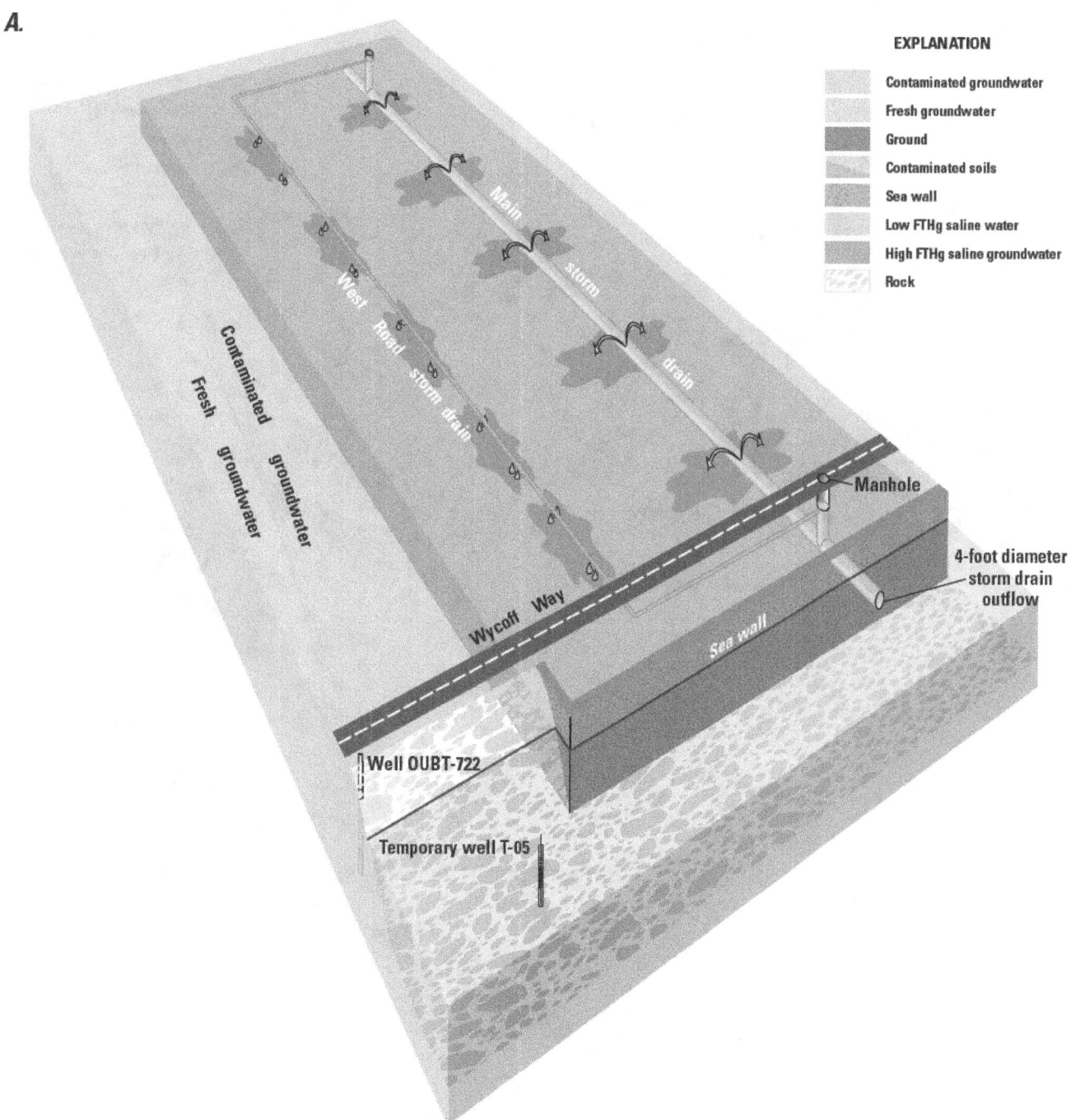

Figure 15. Extraction of Site 2 soils by seawater leaking from and flowing to PSNS015 stormwater drain system on the Bremerton naval complex, Kitsap County, Washington. (*A*) Seawater leaking into soils from the main storm drain and near-surface side drains at high tide and extracting THg from soils. (*B*) Contaminated saline groundwater flowing back into the storm drains and along the seawall to OUBT-722 and piezometerTW-05 from the main storm drain and near-surface side drains at high tide and extracting THg.

B.

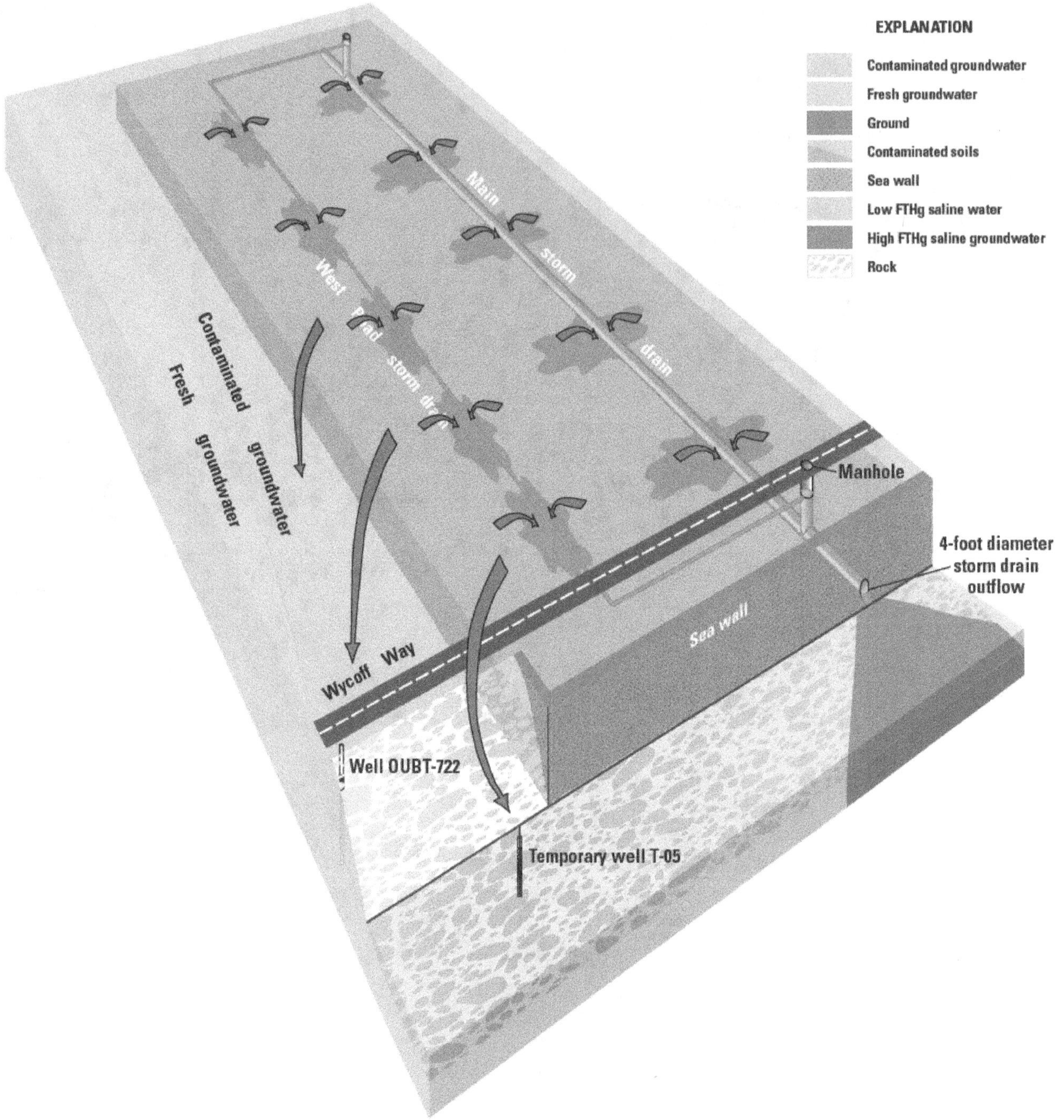

EXPLANATION

Contaminated groundwater
Fresh groundwater
Ground
Contaminated soils
Sea wall
Low FTHg saline water
High FTHg saline groundwater
Rock

Main storm drain

West Brad storm drain

Contaminated groundwater

Fresh groundwater

Wycoff Way

Well OUBT-722

Sea wall

Manhole

4-foot diameter storm drain outflow

Temporary well T-05

Figure 15.—Continued

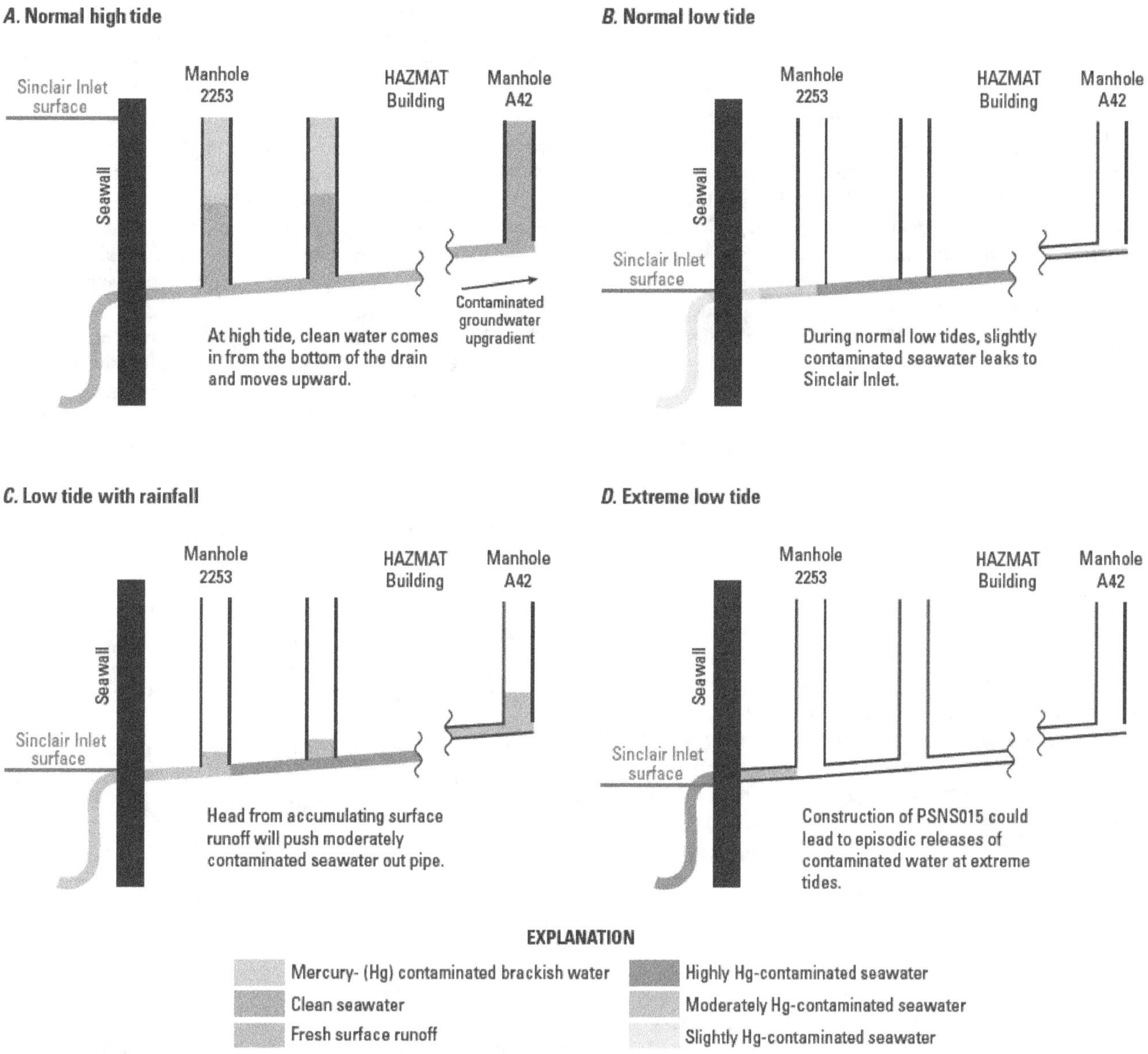

Figure 16. Episodic releases of filtered total mercury during certain tidal conditions. (*A*) Normal high tide, (*B*) normal low tide, (*C*) episodic release during low tide with rainfall, and (*D*) episodic release during extreme low tide.

During extreme low tides of -1 m MWWL (often seen within a month of the summer solstice), the episodic emptying of the PSNS015 stormwater drain pipe in the absence of storm events could cause a large loading of FTHg to Sinclair Inlet (fig. 16D). The highest FTHg concentration measured in marine waters of Sinclair Inlet by USGS (11.4 ng/L) was in a sample collected on June 2, 2009 at station BNC-39 near the end of the submerged PSNS015 pipe. WTHg concentrations as high as 10.71 ng/L were measured in upper layer waters from OU B Marine (table C1) at a station off Site 2.

The residual effect of tidal flushing is hypothesized to involve a complex sequence of physical and geochemical processes that separates the FTHg from the salt. First, FTHg in the contaminated seawater is pushed up drain during the flooding tide. Next, FTHg absorbs onto surfaces, such as the storm drain pipe, vaults and soils upgradient of the contaminated site. During the ebbing tide, the seawater is drained from the system, leaving some FTHg behind. Finally, FTHg is released to lower-FTHg, fresher stormwater that comes in contact with these surfaces.

Capture Zone of Sumps

The stormwater collected from the PSNS&IMF had median FTHg concentrations of 1.83 ng/L. The calculated loading was 0.48 grams of FTHg per year using equation 9 and the flow from table 2, indicating that discharge of FTHg from PSNS&IMF is not a significant source of FTHg to Sinclair Inlet.

Using FTHg concentrations measured by the WMRL, the median FTHg concentration of discharges collected from sump wells of Dry Docks 1-5 results in a loading of 6.9 grams of FTHg per year according to equation 9. The loading varies between 3.2 and 26 grams of FTHg per year depending on whether the non-detectable concentration from the NWQL was assumed to be 0 or 6 ng/L (table 4). The median FTHg concentration collected from the sump well of Dry Dock 6 was 1.81 ng/L, which results in an estimated annual loading of 11.4 grams of FTHg according to equation 9 (table 4). Estimates of FTHg loading from Dry Dock 6 ranged between 8.2 and 13.9 g/yr, depending whether non-detectable concentrations measured by the NWQL were assumed to be 0 or 6 ng/L.

The 18.3 grams of FTHg per year discharged from the dry docks (table 4) originated from FTHg in (1) the seawater seeping into the dry docks, (2) FTHg in ambient groundwater captured by the dry dock sumps, (3) the release of THg from contaminated BNC soils into fresh or saline water that is captured by the sumps, and (4) FTHg inputs from naval operations (table 8). The calculated contribution from ambient groundwater drawn in the sumps was 2.9 grams of FTHg per year based on a concentration of 1.1 ng/L for fresh, deep ambient groundwater collected from well OUNSC 380 (table 5). Using the mean FTHg concentration of 0.31 ng/L for Sinclair Inlet seawater collected in the lower layer OU B Marine (table 3), the calculated contribution from seeping seawater is 2.6 grams of FTHg per year (range: 1.8 to 6.0 grams of FTHg per year). The remaining 13 grams of FTHg per year was contributed as FTHg in process water added by naval operations and from leaching of BNC soils by saline or fresh groundwater that is captured by the dry docks.

Table 8.   Possible sources of filtered total mercury discharged by the dry dock relief drainage systems, Bremerton naval complex, Kitsap County, Washington.

[**Abbreviations:** m^3/s, cubic meter per second; ng/L, nanogram per liter; g/yr, gram per year; NA, not applicable; OU B, Operable Unit B]

	Water (m^3/s)	Freshwater	Number of samples	Concentration (ng/L)		Loading (g/yr)	
				Median	Range	Median	Range
Ambient groundwater	NA	0.083	1	1.1	NA	2.9	NA
Seeping seawater from OU B Marine	0.27	NA	13	0.25 ([1]0.31)	0.21–0.72	2.6	1.8–6.0
Process water	0.010	0.010	NA	NA	NA	NA	NA
Possible sources from contaminated saline groundwater	0.27	NA	8	2.62	1.07–31	2.0	7–260
Total for dry docks	0.36	0.093	NA	NA	NA	18.3	NA

[1]Mean for a large data set of 21 samples.

Only 1 of 13 groundwater samples (13 percent) collected by USGS in the capture zone of the dry docks contained a FTHg concentration greater than 10 ng/L (table 5). A concentration of 31 ng/L was measured at well OUBT-709 (fig. 4), in an area filled with naval waste materials (Site 10C in U.S. Navy, 1992). Additionally, one sample measured by the LTMP in 2005 from LTMP-4 (fig. 3) in another area of fill material (Site 10E in US. Navy, 1992) contained a WTHg concentration of 5,240 ng/L. There clearly is a large range of measured FTHg concentrations in groundwater thought to be captured by the dry dock relief drainage systems (table 8). Equally large is the range of median FTHg concentrations in dry dock discharges. Consequently, no definitive conclusion can be made on the relative importance of naval operations and contaminated groundwater contributing to the FTHg loading from the dry docks.

Marine Sedimentary Sources

Although the FTHg in marine sediment porewater drives the flux of FTHg out of the sediment, the flux of FTHg out of Sinclair Inlet sediment was assessed independently of the porewater concentrations by conducting core incubation experiments as part of the Methylation and Bioaccumulation Project (Huffman and others, 2012) using sediment collected from three stations in greater Sinclair Inlet and three stations in OU B Marine (fig. 4). The highest station median fluxes of FTHg of 269 and 161 ng m^{-2} day^{-1} (table 7) were calculated from increases in FTHg in water over stirred incubated cores collected from Sinclair Inlet stations SI-IN and SI-PO, respectively. The large range of values at each site demonstrates the seasonal variability of the FTHg flux from the sediment. Similar to FTHg porewater concentrations, the largest FTHg fluxes were calculated from incubation studies of cores collected in June or August 2009 and the smallest FTHg fluxes were calculated from cores collected during February 2009. The flux of FTHg from sediment is highly dependent on the season of the year and appears to be controlled by the deposition of organic matter to sediments from biological productivity in the upper layer. The median flux of the seven stations was 46 ng m^{-2} day^{-1}.

It is imprudent to attempt to quantitatively scale the results from these seven stations to a Sinclair Inlet-wide flux from sediment without fully understanding the factors that control the flux both geographically and temporally. To provide an order of magnitude assessment of the release THg from sediment, the minimum (BNC-39) and maximum (SI-IN) median fluxes for stations with at least three flux measurements were applied to the sediment surface area of Sinclair Inlet according to:

$$L_S = \text{Flux} \times A_{SI}, \tag{12}$$

where

L_S is the sedimentary loading,
Flux is the sedimentary flux of FTHg, and
A_{SI} is the area.

This order of magnitude analysis indicates that between 67 and 820 grams of FTHg are added to Sinclair Inlet each year. This estimate can only be refined with a better understanding of the seasonal variation over the entire Sinclair Inlet. Thus, the flux of FTHg from Sinclair Inlet sediment is large and potentially in the hundreds of grams.

Ranking of Sources of Filtered Total Mercury to Sinclair Inlet

Most of the 0.33 ng/L of FTHg in the water column of Sinclair Inlet originates from Puget Sound. The sources of FTHg added directly to Sinclair Inlet were divided into four broad categories based on their mass loading: category (IV), sources greater than 100 grams of FTHg to Sinclair Inlet; category (III), sources ranging between 10 and 100 grams of FTHg to Sinclair Inlet; category (II), sources ranging between 1 and 10 grams of FTHg to Sinclair Inlet; and category (I), sources discharging less than 1 gram of FTHg per year. Category IV sources include diffusion from the Sinclair Inlet sediment and the combined freshwater discharge and tidal flushing from the largest stormwater drain system (PSNS015) on the Bremerton naval complex. However, the numerical values of the loading of these two sources are highly uncertain. The four Category III sources include atmospheric deposition, and discharge from creek basins, greater Sinclair Inlet stormwater basins and the industrial BNC dry dock systems. If FTHg concentrations in streams are higher during winter conditions than summer conditions, these calculations will cause an underestimate of the annual loadings of FTHg. The loading of FTHg in groundwater discharging to greater Sinclair Inlet has not been assessed. It is uncertain if the FTHg being discharged by the dry docks originates from naval operations or extraction of THg from BNC soils as the dry dock sumps draw in ambient fresh groundwater and saline water from Sinclair Inlet. The release of FTHg by groundwater from two fill areas that may be captured by the sumps needs further study. Category II sources include municipal effluent discharging into Sinclair Inlet, industrial discharges from the BNC steam plant, and groundwater water discharge around the end of the seawall in the Vicinity of Site 2. The conversion of the steam plant to a reverse osmosis system in 2010 likely decreased this FTHg source. Stormwater discharges from the Zone of Direct Discharge and Capture Zone of the Sumps do not appear to be a significant source of FTHg (category I).

Concentrations of Total Mercury of Solids Discharged to Sinclair Inlet

In the absence of a balance between the mass of solids entering the water column of Sinclair Inlet and the mass of solids leaving the water column through sedimentation or advection out of Sinclair Inlet (table 2), a qualitative analysis of the source of solid-phase THg is presented here. This qualitative analysis focuses on comparing THg concentrations of suspended solids among source waters (tables C2–C3) and with solids suspended in the water column of greater Sinclair Inlet and OU B Marine (table C4). Although this analysis is not a comprehensive mass balance on solid-phase THg, it does provide insight into critical issues for the Sinclair Inlet concerning how quickly (if at all) natural recovery is currently proceeding, and whether there is a need for additional remediation of mercury sources.

Solids Discharged to Greater Sinclair Inlet

Solids containing THg are discharged to greater Sinclair from creek and stormwater basins and from wastewater treatment plants. This PTHg originates from dry fall on the landscape of Sinclair Inlet watershed, the sequestering of FTHg in rainwater by soil particles, leaching of THg from natural soils, and use of mercury by industry, commerce, and households within the basin. Although the average THg dry deposition of 0.65 ± 0.11 μg m^{-2} yr^{-1} was measured in dry fall by Brandenberger and others (2010), corresponding THg concentrations on the deposited particles are not known because TSS concentrations in the samples washed off the sampler were not measured.

Creek

THg concentrations of suspended solids in four creeks in May and July 2008 were generally less than 0.25 mg/kg (fig. 17), except for the two samples from Annapolis Creek that averaged 0.27 mg/kg (table C2). The maximum THg concentrations of solids inferred from the slope of WTHg versus TSS concentrations (table A7) measured by ENVVEST (0.064 to 0.17 mg/kg) were similar to THg concentrations of solids measured by USGS (0.064 to 0.23 mg/kg). The THg concentrations of solids from Olney Creek (0.14 mg/kg), just outside of the model domain, were within the range of THg of solids for the four Sinclair Inlet creeks.

Stormwater Basins

Concentrations of THg of solids measured on stormwater particles discharged into Sinclair Inlet (table C2) were 0.096 and 0.138 mg/kg and were slightly less than the measured THg concentration of solids (0.157 mg/kg) from a City of

Bremerton storm drain discharging outside of Sinclair Inlet (Sheridan Road site in fig. 1). The maximum concentration of solids inferred from the slope of WTHg versus TSS measured by ENVVEST for samples collected at location LMK122 (0.38 mg/kg) was greater than the THg concentration of solids (0.096 mg/kg) collected by USGS at the same location (Navy City near Gorst in table C2). The USGS sample may have been affected by road dirt from a nearby highway construction site flowing into the drainage ditch during storm sampling.

Wastewater Effluents

THg concentrations of wastewater solids filtered from the City of Bremerton wastewater treatment plant (WWTP) effluent were measured from three 5-L composite samples prepared in May, July, and August 2009. THg of wastewater solids ranged from 0.183 to 0.210 mg/kg (table C2 and fig. 17).

Solids Discharged from the Bremerton Naval Complex

Sources of PTHg discharged from the BNC include industrial sources and stormwater discharge. USGS sampled water discharged from four industrial outfalls from three regulated sources that included sump water from two dry dock relief drainage systems and the industrial effluent from a steam plant during 2007 and 2008. As with FTHg, the release of PTHg from tidal flushing of the PSNS015 stormwater drain system also was examined. Although transport of PTHg by groundwater discharge to Sinclair Inlet is not included in the mass balance, solids in ambient and BNC groundwater are captured by the dry docks sumps; and sump water is discharged to Sinclair Inlet.

Zone of Direct Discharge

TSS concentrations (0.75–1.45 ng/L) and TSS loading (0.09 metric tons) in steam plant effluent were low. Concentrations of THg of solids varied from 2.95 to 68.74 mg/kg (fig. 18, table C3). Similar to FTHg in steam plant effluent, THg of solids increased with increasing specific conductance (fig. 9). Similar to FTHg, THg concentrations of steam plant solids likely decreased with conversion to a reverse osmosis treatment system in 2010.

Vicinity of Site 2

A TSS concentration of 222 ng/L was measured in non-saline stormwater collected from stormwater drain PSNS015 within Site 2 during the January 9, 2009, storm event. The THg concentration of solids in this non-saline water was 1.49 mg/kg (fig. 18).

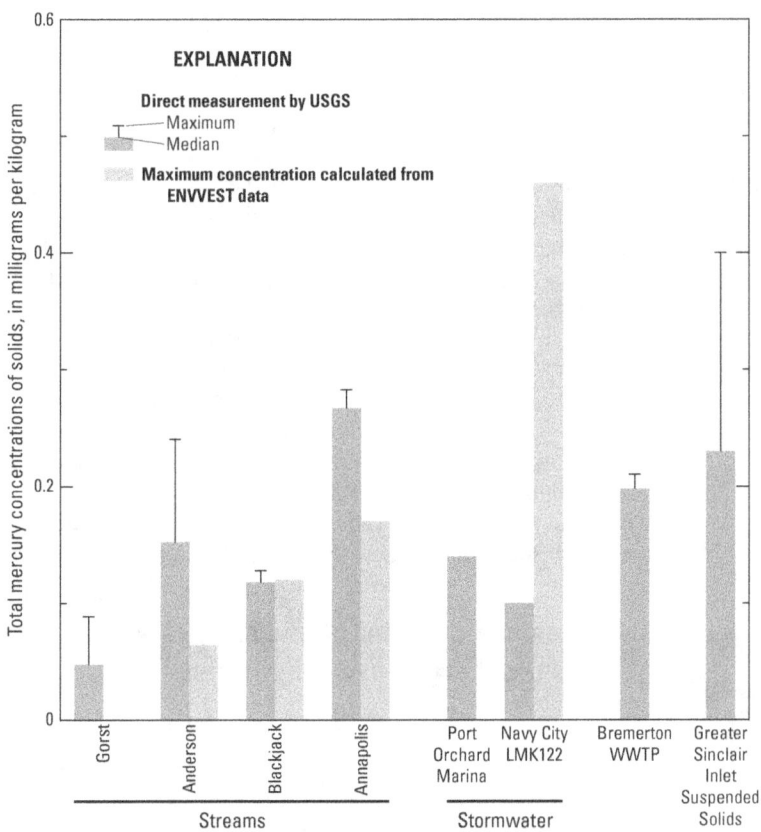

Figure 17. Concentrations of total mercury of suspended solids in streams, stormwater, and wastewater treatment plants in the Sinclair Inlet basin, Kitsap County, Washington measured by USGS (2008–09), maximum concentrations calculated from ENVVEST data (2002–05), and concentrations in solids suspended in the upper layer of greater Sinclair Inlet.

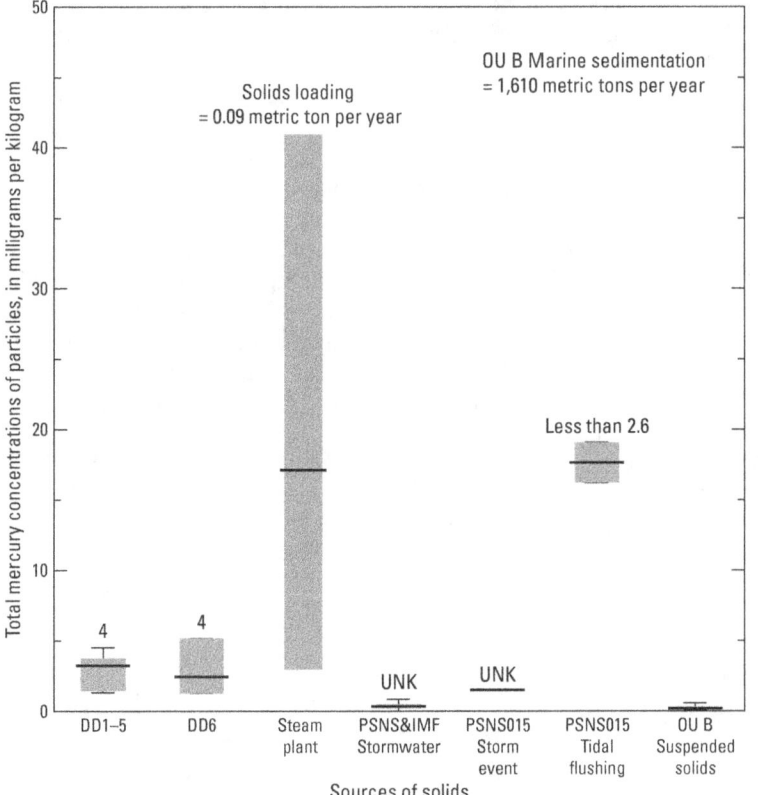

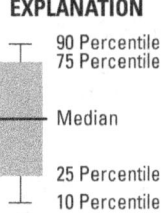

Figure 18. Concentrations of total mercury of solids originating from Bremerton naval complex, Kitsap County, Washington. (The whisker box plots do not include samples with total suspended solid concentrations less than 0.5 mg/L that likely are biased by sequential sampling during changing TSS conditions. The labels above the whisker boxes indicated the estimated loading of total suspended solids in metric tons per year. UNK indicates solids loading is unknown.)

The THg concentrations of suspended solids for the two sampling events of the ebbing tidal cycles on December 29, 2010, and March 31, 2011, were 16.2 and 19.1 mg/kg, respectively (fig. 16 and table C3). The THg concentrations of these solids were considerably larger than the THg concentrations of solids in the PSNS015 stormwater drain associated with the storm event (1.49 mg/kg), but within the range of THg concentrations found for soils in Site 2 (table A1). Thus, the small amount of solids flowing from PSNS015 stormwater drain during ebbing tides are distinctly different than storm drain particles from surface runoff and the solids discharged during non-storm ebbing tides appear to be soil particles leaking into the PSNS015 stormwater drain system.

Capture Zone of the Sumps

The loading of solids from BNC stormwater drains is unknown. Two stormwater samples collected from PSNS&IMF during storm events contained high concentrations of TSS (table C3). The concentrations of THg of the stormwater solids were 0.21 and 0.495 mg/kg (fig. 18).

TSS concentrations in dry dock discharges ranged from 0.16 to 1.47 mg/L, except when maintenance of the sump wells was occurring (June 24, 2008). When TSS concentrations were greater than 0.5 mg/L, THg concentrations of solids were less than 6 mg/kg, with median concentrations of 3.23 mg/kg and 2.42 mg/kg for solids from Dry Docks 1–5 and 6, respectively (table C3). The highest THg concentrations of solids collected from the Dry Docks 1–5 system (13.2 mg/kg in December 2007) and from the Dry Dock 6 system (17.7 mg/kg in April 2008) were associated with the lowest TSS concentrations for each dry dock system (0.49 mg/L and 0.16 mg/L, respectively).

Concentrations of metals increase with decreasing particle size because of increasing surface-to-volume ratios. Physical settling of the coarser fraction of the original particles entering the dry dock drainage system is a likely mechanism causing the observed relation between mercury concentrations of solids versus TSS concentrations. Particle-size effects may have been exacerbated by the sequential sampling for PTHg and TSS during the pumping cycle. When the pumps turn on, solids near the pump inlet likely are resuspended by turbulence, resulting in higher TSS and PTHg concentrations at the beginning of the pumping cycle when the PTHg samples were collected. As the pumping cycle continues, the water that is being pumped comes from the increasing distances from the

pump, which likely had lower TSS concentrations. Since the TSS sample was always collected last, the TSS sample likely underestimates the TSS concentration of water at the time particulate THg samples were collected at the beginning of the cycle.

For the December 2007 sampling of Dry Dock 1-5 from the discharge pipe of Pump 4, the dramatic difference between the THg concentration of solids calculated from sequential sampling with two different bottles and the THg concentration of solids calculated from PTHg and TSS samples collected from a single bottle containing the 24-hour composite sample support the hypothesis of bias from sequential sampling during changing TSS conditions. The THg concentration for the sequential sampling of two grab samples at the Dry Dock 1–5 system was 13.2 mg/kg. In contrast, the THg concentration of solids obtained from PTHg and TSS measurements aliquots from the same composite bottle filled the previous 24 hours was near the median concentration (3.75 mg/kg).

Synthesis of Concentrations of Total Mercury of Solids Discharged to Sinclair Inlet

Solids discharged into greater Sinclair Inlet and OU B Marine enter the water column after undergoing geochemical processes occurring in brackish water and physical settling of coarser particles in the small mixing zone of the discharge. Thus, source solids contribute to the population of solids suspended in the upper layer of Sinclair Inlet, but the exact contribution of the sampled solids is unknown because of the lack of a mass balance of solids for Sinclair Inlet. THg concentrations of solids (fig. 17) from streams (0.64–0.283 mg/kg), stormwater (0.96–0.157 mg/kg), and even wastewater effluent (0.183–0.210 mg/kg) discharging into greater Sinclair Inlet generally were within the range of concentrations of THg of solids suspended in the upper water column of greater Sinclair Inlet (0.23+0.17 in table 3). Likewise, THg concentrations of solids (fig. 18) from the PSNS&IMF stormwater drains (0.21–0.495 mg/kg) were within the range of THg concentrations of solids suspended in the upper layer of OU B Marine (0.23+0.13 in table 3).

In contrast, THg concentrations of solids from the steam plant (2.95–40.9 mg/kg) and from stormwater drain PSNS015 during tidal flushing (16.2–19.1 mg/kg) were significantly higher than THg concentrations of solids suspended in the upper water column of OU B Marine (fig. 18). The THg concentrations of particles discharged from the dry dock drainage relief systems (1.32–5.17 mg/kg) and from the

PSNS015 stormwater drain during a storm event (1.46 mg/kg) were slightly higher than THg concentrations in OU B Marine. Statistical comparisons were not performed on the source solids because it is not known how the collected particles are representative of the population of solids in either the water column or the sediment in the absence of a mass balance of solids.

Direct settling of suspended solids within the water column of this U-shaped basin (fig. 6) has been measured (U.S. Navy, 2007b cited in Paulson and others, 2010). This mechanistically direct transport of solids between the water column and the sediment column is examined by comparing THg concentrations of suspended in the water column of Sinclair Inlet with STHg concentrations of bulk Sinclair Inlet sediment. The quality of the STHg measurements of whole sediment samples by the LTMP was assessed in Paulson and others (2010) and found to be acceptable. The percentages of fines of the whole LTMP sediment are shown in figures 2 and 3. Additionally, the large number of measurements of THg of suspended solids from the marine water column in Sinclair Inlet (table 3) by USGS as part of the Methylation and Bioaccumulation Project in 2008 and 2009 allows statistical comparison with STHg concentrations reported by the U.S. Navy between 2003 and 2007, and in 2010 as part of the LTMP. In this study, the cumulative probability distributions of THg concentrations on suspended solids and of STHg concentrations of sediment samples are compared separately for OU B-Marine and greater Sinclair Inlet. Cumulative probability distributions provide not only information on the median (a cumulative probability of 0.5), but also provide a visual depiction of distribution of concentrations.

The cumulative probability distribution of THg concentrations of solids suspended in the upper layer of greater Sinclair Inlet (fig. 19) is statistically ($p < 0.001$) lower than (to the left of) the distribution of STHg concentrations measured before this study (2003–07). This observation indicates that sedimentation of these lower concentration THg-solids in the water column should cause a decrease in STHg concentrations in surface layer of Sinclair Inlet sediment over time. This decrease would occur even if preferential settling of coarser, larger-sized solids were occurring in the water column. The coarser, larger-sized solids likely would have a lower THg concentration because of the lower surface area-to-volume ratio of these larger-sized solids compared to the smaller-sized solids that might be advected to

Puget Sound. THg concentrations of solids suspended in the lower layer of the water column were not statistical different from THg in the upper water column (both category B for the post-Anova Tukey multicomparison test in last column of table 3). The cumulative probability distribution of STHg in Sinclair Inlet sediment collected after the study in 2010 (U.S. Navy, 2011) was to the left (lower concentrations) of that of sediment collected in 2003, 2005, and 2007, but not statistically lower (p value=0.18). A rate of decrease of STHg concentrations in sediment of Sinclair Inlet cannot be accurately modeled until the sources of the solids suspended in Sinclair Inlet are quantified.

Similar to greater Sinclair Inlet, THg concentrations of solids suspended in the upper layer of OU B Marine collected (table 3) in 2008 and 2009 were significantly lower ($p < 0.001$) than STHg concentrations of sediment from OU B Marine collected before the study (fig. 20). Unlike greater Sinclair Inlet, THg of solids in the bottom layer of OU B Marine (multicomparison category A in table 3) were significantly greater than the upper layer of OU B Marine (category B). The resuspension of higher THg-sediment sediment was probably increasing the overall THg concentration of suspended solids present in the bottom layer. Settling of the suspended solids and deposition to OU B Marine sediment column should decrease STHg concentrations. The cumulative probability distribution of STHg of sediment collected after the study (2010) was significantly lower (p value=0.015) than that for sediments collected before the study. In the case of OU B Marine sediment, the difference in the cumulative distribution before and after the USGS study is confined to the trend of lower STHg concentration only above the median STHg concentration (50 percent on the cumulative probability distribution).

Local settling of solids with high THg-concentrations solids discharged from several BNC sources could reverse the general positive effects of settling of low-THg concentration suspended solids in the water column. For instance, both the total organic carbon content and the STHg concentrations in grid cell 39, the cell in which stormwater drain PSNS015 solids are discharged, increased two-fold or more between 2007 and 2010. The effect of these BNC sources of PTHg on THg sedimentation can be assessed when THg concentrations of dry dock solids not affected by bias of sequential sampling during changing TSS conditions and of steam plant solids from the current reverse-osmosis system are obtained.

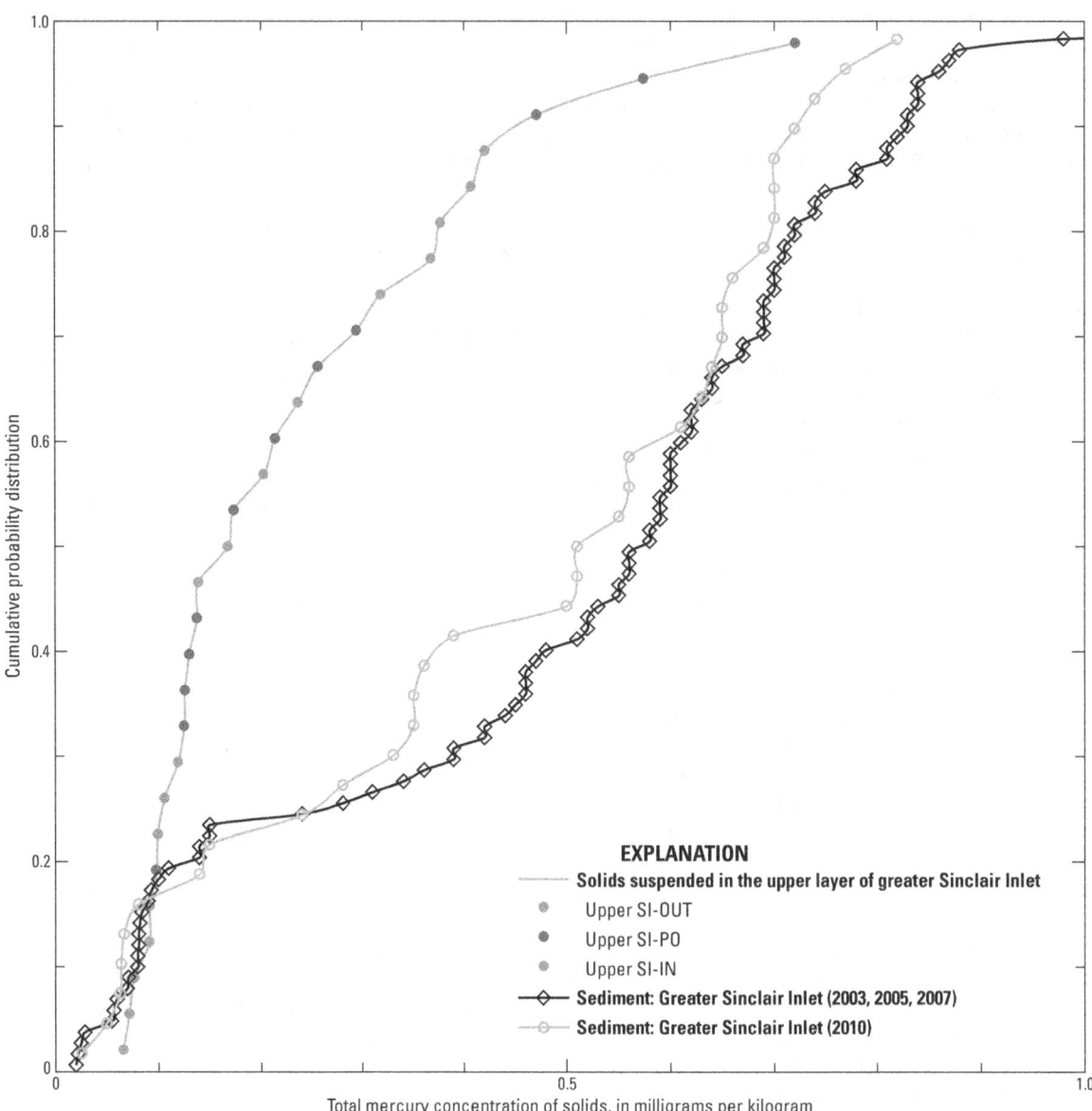

Figure 19. Cumulative probability distribution of total mercury concentrations of solids suspended in the upper layer of the water column, of surface sediment (2003–07) and of surface sediment (2010) from greater Sinclair Inlet, Kitsap County, Washington.

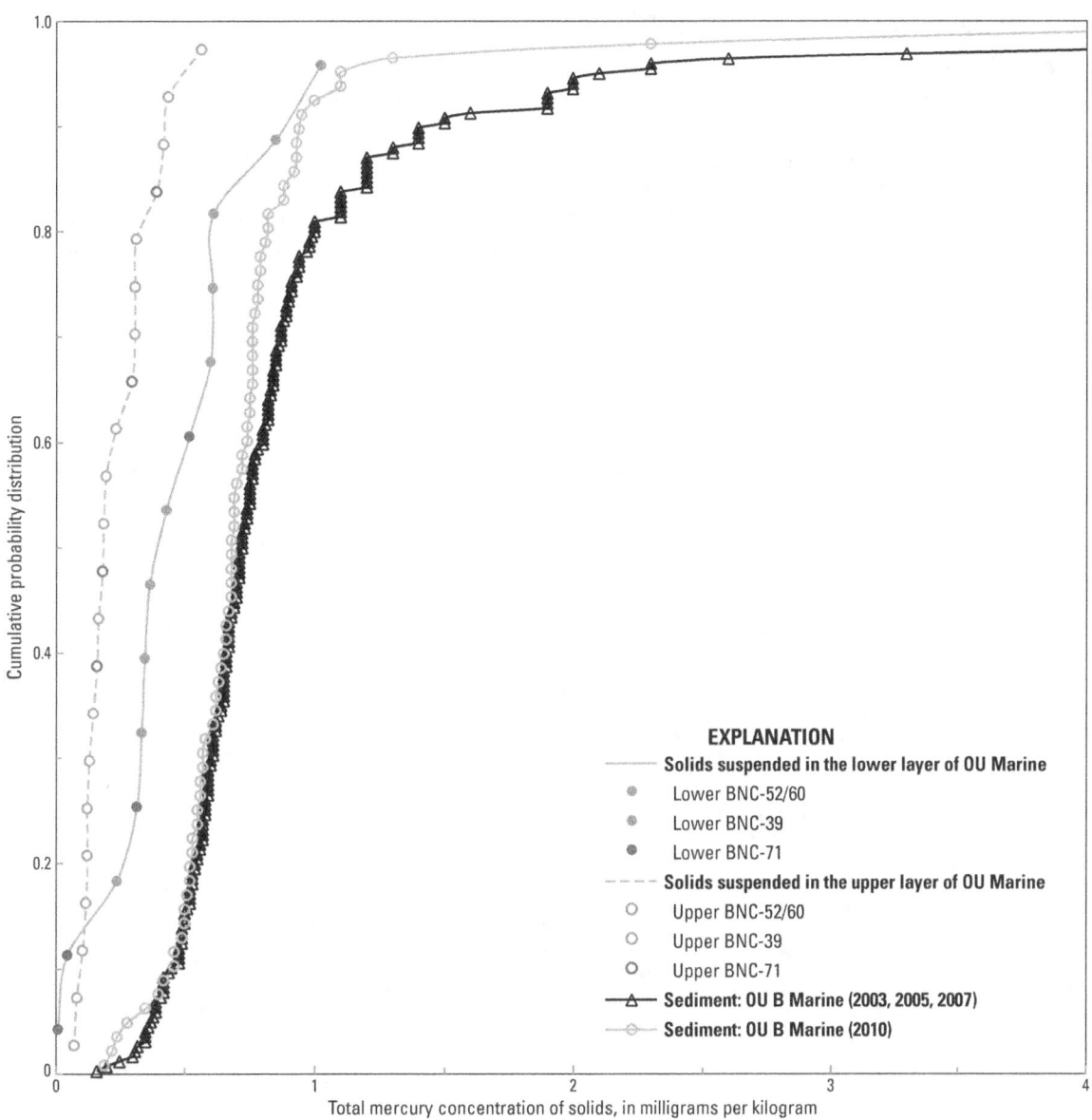

Figure 20. Cumulative probability distribution of concentrations of total mercury of solids suspended in the upper layer of the marine water column (2008–09) and total mercury in sediment collected between 2003 and 2007, and in 2010 from OU B Marine of the Bremerton naval complex, Kitsap County, Washington.

Methylmercury in the Sinclair Inlet Drainage Basin

Methylmercury measurements were conducted as a preliminary survey of the occurrence and concentrations of FMHg in water sources to Sinclair Inlet, and therefore, are not suitable for a comprehensive mass balance of FMHg in Sinclair Inlet.

Greater Sinclair Inlet

FMHg was detected by USGS in filtered freshwater from three storm drains (Sheridan Road in fig. 1, and Navy City and PO-Boat ramp in fig. 4) discharging outside of the BNC at a concentration of 0.06 ng/L (fig. 21). The median FMHg concentrations of six filtered effluent samples from the two wastewater treatment plants was 0.12 ng/L, ranging from 0.05 to 0.21 ng/L. A non-parametric correlation

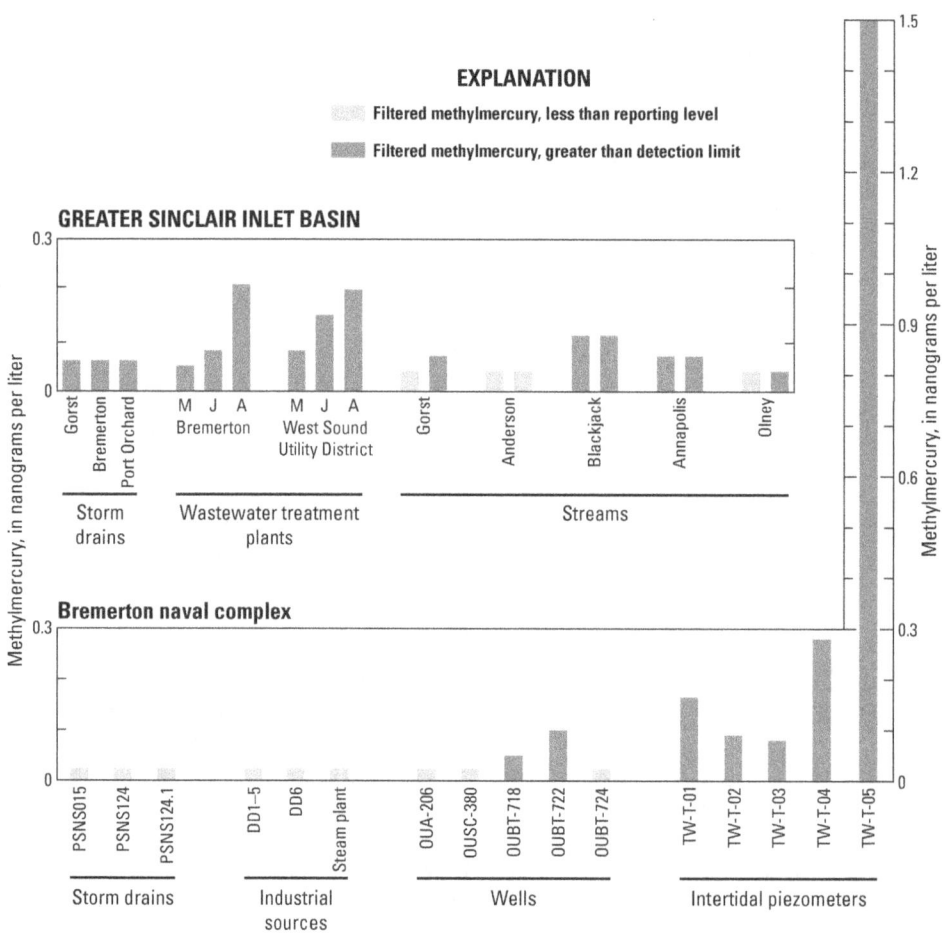

Figure 21.   Methylmercury concentrations in terrestrial sources within the Bremerton naval complex (lower) and from basins of greater Sinclair Inlet (upper), Kitsap County, Washington. Note that the two bar graphs are the same scale, but the scale for intertidal piezometersTW-T05 extends to 1.5 ng/L. The three bars for the wastewater treatment represent data collected by USGS in May (M), July (J), and August (A) 2009.

analysis (Kendall tau) was performed using redox-sensitive constituents (DOC, ammonia, nitrite, iron and manganese) as independent variables. These reduced biogeochemicals are indicative of the chemically-reduced conditions that are generally associated with biogeochemical transformation of mercury to methylmercury (Marvin-DiPasquale and Agee, 2003). The correlation analysis indicated that neither the FMHg concentration nor the ratio of FMHg to FTHg was significantly correlated with these constituents. In the contrast, the consistent increase of FMHg concentrations in wastewater effluent samples from May to July to August, 2009, strongly suggests that ambient temperature may be the primary factor controlling MHg concentration in such waters (Gbondo-Tugbawa and others, 2010).

Water from Gorst, Anderson, Blackjack, Annapolis and Olney Creeks (fig. 5) were sampled in May and July 2008. In 6 of the 10 filtered creek samples FMHg was measured at concentrations ranging from 0.05 to 0.11 ng/L (fig. 21). In contrast to wastewater effluents, the non-parametric correlation analysis of the results of the 10 stream samples (table 9) indicates that filtered total iron concentrations are significantly correlated with FMHg concentrations (p value = 0.029), as was observed by Balogh and others (2004). However, the finding that FMHg is not correlated with DOC (p = 0.11) is contradictory to a number of studies that have found correlations between MHg and DOC in freshwaters (Babiarz and others, 2001; Shanley and others, 2002). In contrast, the ratio of FMHg to FTHg was correlated with DOC (p value = 0.029), filtered nitrite (p value = 0.047), total iron (p value = 0.005), and total manganese (p value = 0.029).

Bremerton Naval Complex

FMHg was not detected in most of the sources of water discharging to Sinclair Inlet from the BNC. FMHg was not detected at a reporting level of 0.04 ng/L in water discharged from six BNC outfall systems including sump water from two dry dock relief drainage systems; the industrial effluent from a steam plant sampled during April 2008; and three stormwater drains sampled on January 7, 2009 (fig. 21).

FMHg was only detected in groundwater from the Vicinity of Site 2 and was not detected at a reporting level of 0.04 ng/L in groundwater from the Zone of Direct Discharge and the Capture Zone of the Sumps collected during April 2008. The estimated FMHg concentration in fresh groundwater from well OUBT-718 (0.05 ng/L) was slightly above the reporting level. The detection of FMHg in well OUBT-718 was not associated with an elevated

Table 9. Non-parametric correlation statistics of filtered methylmercury concentrations relative to concentrations of reduced biogeochemicals.

Reduced biogeochemical	Kendall Tau correlation coeffecient (p value)	
	Filtered methylmercury	Ratio of filtered methylmercury to filtered total mercury
Organic carbon	0.48 (0.11)	0.54 (0.029)
Ammonia	0.08 (0.86)	0.07 (0.86)
Nitrate plus nitrite	-0.03 (1.00)	0.18 (0.48)
Nitrite	0.34 (0.29)	0.55 (0.047)
Iron	0.61 (0.029)	0.70 (0.005)
Manganese	0.43 (0.16)	0.54 (0.029)

THg concentration (0.8 ng/L), but it was associated with reducing conditions inferred from the presence of reduced species including hydrogen sulfide, ferrous iron, filtered total manganese, and filtered ammonia. In contrast to well OUBT-718, groundwater from well OUBT-722 was saline and oxygenated. FMHg in groundwater from well OUBT-722 was 0.10 ng/L, whereas the FTHg concentration was 581 ng/L.

FMHg was measured once in saline water collected from the five piezometers monitored during the intertidal study in June 2009 (fig. 5). FMHg concentrations in the four westernmost piezometers ranged from 0.09 to 0.28 ng/L, even though THg concentrations in these four samples ranged from 0.21 to 2.30 ng/L. The easternmost piezometer (TW-T05) near the seawall contained the highest FMHg concentrations of groundwater found on the BNC (1.5 ng/L) with a corresponding FTHg concentration of 244 ng/L. Although no chemical data on redox-sensitive species were collected during the intertidal study, anecdotal observations indicate that reduced Fe was present in groundwater from this piezometer because: (1) the particles collected on the two filters from this piezometer were bright red-orange and (2) the filtered, unacidified samples delivered to the laboratory had a strong reddish color several hours after filtration that disappeared upon acidification. These two observations suggest the presence of ferric iron from the oxidation of reduced ferrous iron formed in the soils of the intertidal zone. Thus, the chemistry of water collected from piezometer TW-T05 resembled the chemistry of porewater from reduced marine sediments of Sinclair Inlet (Huffman and others, 2012) more than the oxygenated groundwater found throughout most of the BNC.

Sinclair Inlet Sediment

FMHg concentrations in 5 of the 24 porewater samples collected from OU B Marine and greater Sinclair Inlet exceeded 10 ng/L. The maximum FMHg concentration of 28 ng/L (table 10) was measured at marine sediment station SI-IN in August 2008 . Median FMHg concentrations for porewaters collected from each OU B Marine station ranged from 0.09 to 1.1 ng/L. In contrast, the median FMHg concentrations for stations SI-IN and SI-PO were about 10 ng/L. The median FMHg concentration of porewaters from the open water SI-OUT was 0.10 ng/L.

Synthesis of Methylmercury in Sinclair Inlet

FMHg was generally not detected or measured at concentrations less than 3 times the reporting level (0.04 ng/L) in fresh surface water discharging into the Sinclair Inlet. Of the industrial discharges, stormwater samples, and groundwater samples collected within the BNC, FMHg was only found in groundwater in the Vicinity of Site 2 and was associated more with reducing conditions in non-saline groundwater or saline intertidal water than with high FTHg concentrations. The highest concentration of FMHg was collected in groundwater that appeared to contain reduced iron. The highest median FMHg concentration for a source category was measured in wastewater-treatment plant effluent (0.12 ng/L). The source category with the next highest median FMHg concentration was rainwater (0.08 ng/L). The median concentration of FMHg in creek water was 0.06 ng/L. Of the five streams, Blackjack Creek had the highest FMHg concentration (0.11 ng/L). The fraction of FTHg in the FMHg form in creek water was more highly correlated with reduced biogeochemicals than were the absolute concentrations of FMHg. The highest FMHg concentrations of the study were detected in porewater of highly reducing sediment collected from OU B Marine and greater Sinclair Inlet.

Observations Indicating the Need for Further Study

To fully assess present-day sources of mercury to Sinclair Inlet, the need to fill significant data gaps became apparent during the course of this project. In November 2011, the U.S. Navy requested that the USGS collect additional data to fill these data gaps. Sample collection was completed in March 2012. When laboratory analyses have been received and the data have been synthesized, a supplement to this report that includes additions and revisions to this report (for example, table 4) will be published as a separate document.

Table 10. Concentrations of filtered methylmercury in porewaters from three stations in the Bremerton naval complex and three stations in greater Sinclair Inlet, Kitsap County, Washington, August 2008–August 2009.

[Numbers after BNC stations refer to the grid cell shown in figure 5. **Abbreviations:** FMHg, filtered methylmercury; ng/L, nanogram per liter; SI, Sinclair Inlet; OUT, Outer; IN, Inner; PO, Port Orchard; BNC, Bremerton naval complex; –, not applicable]

Field identifier	Median	Range
FMHg porewater concentrations, ng/L		
SI-OUT	0.10	0.08–0.58
SI-IN	10.18	0.1–28
SI-PO	10.40	0.07–20
BNC-39	0.33	0.14–0.63
BNC-52	0.09	–
BNC-60	1.1	0.24–15
BNC-71	0.88	0.09–1.5

Groundwater

Well OUBT-722 was constructed in fill material adjacent to Site 2 of the Initial Assessment Study (URS Consultants, Inc., 1991). Likewise, wells OUBT-709 and OUBT-724 were constructed in fill material indentified as Site 10 C and Site 10 E in the Site Inspection (U.S. Navy, 1992). Similar to the physical conditions at Site 2, tidal flushing of seawater through the storm drains around OUBT-709 and OUBT-724 may facilitate the release of mercury from fill material, albeit at a lower rate than the much larger and deeper PSNS015 storm drain. If the elevations of storm drains around wells OUBT-709 and OUBT-724 are higher than the elevations of the PSNS015 stormwater drain system, then release from Site 10 C and Site 10 E would occur only during the highest tidal elevations. Two observations related to wells OUBT-724 and OUBT-709 suggest occasional release of THg may be occurring from fill material other than that from than Site 2. The high WTHg concentration of 5,240 ng/L in groundwater from OUBT-724 collected by the LTMP on October 18, 2005 appears to have been collected at the highest tidal elevation of all samples collected from this well. However, the U.S. Navy reported no other WTHg concentration that was acceptable according quality-control criteria for a detection limit of 200 ng/L through October 2007. Using more sensitive analytical measurements, WTHg concentrations of 41.6 and 27.5 ng/L were reported by the LTMP in 2008 and 2009, and this study reported FTHg concentrations of less than 10 and 6.52 ng/L in 2008.

Other than groundwater collected from OUBT-722, the highest FTHg concentration in groundwater collected on the BNC during this study (31 ng/L) was collected from OUBT-709 during winter when groundwater elevations were higher than those during the April sampling (FTHg of 5.76 ng/L). Thus, it is possible that WTHg and FTHg concentrations in OU B Terrestrial wells may be related to the difference between tidal elevation and the elevation of storm drains servicing adjacent paved areas.

Industrial Sources

The highest concentrations of THg of solids discharged from the dry docks sumps were measured in samples with the low TSS concentrations. However, this observation might be an artifact of changing TSS concentrations during the pumping cycle. If the release of a small loading of solids with a higher THg concentration on the particles is of concern, several measurements of THg concentrations of solids can be collected over the pumping cycle by collecting particulate THg and TSS from the same sample container.

The highest concentration of filtered THg in a permitted source was found in the effluent from the steam plant. However, loadings calculations of THg reported in this study probably do not reflect the current loadings of THg from the steam plant operating with the new reverse osmosis demineralization plant. Re-sampling of effluent from the steam plant would provide a more representative estimate of current THg loadings from the steam plant.

Streams

Concentrations of FTHg and THg of solids in five creeks used in the mass loading calculations were measured during base flow conditions in May and July 2008. FTHg concentrations in creek water were significantly less than THg concentrations in rainfall indicating that the watersheds retained most of the THg falling on the landscape. During the flow conditions in wetter or storm conditions when most of the discharge occurs, more overland flow also occurs. Thus, less THg in rainwater is retained by watersheds than that retained during base flow conditions. Collecting FTHg and THg of solids during wet weather conditions will provide a better estimate of annual THg loadings to greater Sinclair Inlet.

Summary

A mass balance of any constituent in an estuary requires simultaneously balancing water, salt, and solids for the estuary. Previous data on the salinity distribution and long-term flows in Sinclair Inlet allow the 98 cubic meter per second estimate of net flux of water entering Sinclair Inlet from Puget Sound to be constrained to within a factor of two. In contrast, the sources of most of solids depositing to sediment of Sinclair Inlet are unknown. In order to maximize the use of previously collected data and data collected by U.S. Geological Survey between 2007 and 2010, a quantitative mass balance of filtered total mercury (FTHg) was conducted along with a qualitative comparison of total mercury (THg) concentrations of various suspended solids and of marine sediment. A preliminary survey of FMHg concentrations in water sources to Sinclair Inlet was also conducted. The ability to predict the recovery of mercury in Sinclair Inlet sediment (whether a source of solids is likely increasing or decreasing THg concentrations in the upper layer of the marine sediments) requires additional information on the physical oceanography and sediment transport of Sinclair Inlet.

Most FTHg in Sinclair Inlet originates from salt water flowing from Puget Sound. Atmospheric, terrestrial, and sedimentary sources contribute approximately 420 grams of FTHg per year and increase FTHg concentrations in Sinclair Inlet to 0.33 nanograms per liter from the 0.2 nanograms per liter in Puget Sound seawater entering Sinclair Inlet. The two major sources of FTHg within the Sinclair Inlet watershed are (1) diffusion from marine sediment and (2) discharge from the largest stormwater drain systems on the Bremerton naval complex (BNC) that pass through a zone of contaminated soil, albeit numerical values of these sources are highly uncertain. The category of sources with second largest loadings of FTHg include (1) THg in rain falling directly on Sinclair Inlet, (2) discharge from creek, and (3) discharge from stormwater basins outside of the BNC. Much of the THg falling in rainwater on the creek basin is retained by the creek basins. A municipal wastewater treatment plant, the effluent from the steam plant on the naval base, and the discharges from two naval dry dock systems constitute the category of the third largest loadings for individual sources. Stormwater discharged by the shipyard and groundwater discharge from the base do not seem to be significant sources of FTHg, although conflicting data prompted additional studies in 2011.

THg concentrations of solids from streams, stormwater, and wastewater effluent discharging into greater Sinclair Inlet were similar to THg concentrations of solids suspended in the upper layer of greater Sinclair Inlet. In contrast, four sources of solids (tidal flushing, stormwater discharge of Puget Sound Naval Shipyard, discharge from the dry docks and discharge from the steam plant) that discharged less than 1 percent of the solids depositing within Operable Unit (OU) B Marine sediments had THg concentrations greater than THg concentrations of solids suspended in the water column. The settling of these high-THg concentrations could be inhibiting the recovery of the total mercury concentration of sediment (STHg) within OU B Marine. The cumulative probably distribution of STHg in sediment of OU B Marine moved to lower concentrations in 2010 compared to 2003–07 and indicated settling of solids from the water column is decreasing STHg concentrations, especially in sediment with the highest STHg. In contrast, sediment near the discharge point of the PSNS015 storm drain showed the highest increase in STHg within the OU B Marine in 2010. This increase near the largest source of mercury from the BNC deserves further attention.

FMHg was detected at low concentrations in storm drains, creeks, and wastewater effluent discharging to greater Sinclair Inlet. Temperature seemed to be controlling FMHg in wastewater effluent, while reduced ferrous iron was associated with FMHg concentrations in streams. FMHg was not detected in any of the storm drains or industrial BNC sources discharging to Sinclair Inlet. The highest FMHg concentrations were associated with porewaters in highly reducing sediment and intertidal groundwater.

Former Site 2 is the major source of mercury discharged from the BNC to Sinclair Inlet, as previous U.S. Navy groundwater data from well LTMP-3 and previous ENVironmental inVESTment program stormwater drain data from PSNS015 have suggested. The site contributes a major fraction of the FTHg added to Sinclair Inlet from atmospheric, terrestrial, and sedimentary sources. Tidal flushing of large amounts of seawater into the PSNS015 storm drain with some subsequent leakage into the aquifer facilitates the extraction of THg from contaminated soils. FTHg is then released to Sinclair Inlet through the intertidal area at the end of the seawall and from the storm drain. A subterranean estuary in the stormwater drain retains THg in the storm drain until the stratification in the subterranean estuary destabilizes and possibly causes episodic releases of FTHg. High concentrations of THg of solids also are being released during tidal flushing and could be the cause of the increasing STHg concentration in adjacent marine sediments. During the final stages of this investigation, a separate release mechanism of THg associated with freshwater was detected associated with is this and needs further evaluation.

Acknowledgments

The authors would like to thank Patric Coxon and the staff of the City of Bremerton Wastewater Treatment Plant and Lance Hunt and the staff of the West Sound Utility District Wastewater Treatment Plant for collection of final wastewater effluent. Current and former U.S. Geological Survey Washington Water Science Center (WAWSC) staff Brietta Carter, James R. Foreman, Sarah A. Henneford, and Karen L. Payne assisted in the sampling of stream and stormwater. Current and former U.S. Geological Survey WAWSC staff James R. Foreman, Morgan E. Keys, Cg Laird, Karen L. Payne, Kelly L. Scholting, and Gary L. Turney assisted in sampling groundwater and industrial sources at BNC. Current and former U.S. Geological Survey WAWSC staff James R. Foreman and John Sottilare; and Wisconsin Mercury Research Laboratory (Middleton, WI) staff David P. Krabbenhoft, Jacob, M. Ogorek, and Thomas G. Sabin

assisted in the collection of marine water column samples. U.S. Geological Survey WAWSC staff James R. Foreman, Cg Laird, and Richard W. Sheibley, U.S. Geological Survey National Research Program (Menlo Park, CA) staff Mark Marvin-DiPasquale and Jennifer L. Agee, and Wisconsin Mercury Research Laboratory (Middleton, WI) staff David P. Krabbenhoft Jacob, M. Ogorek and Thomas G. Sabin assisted in collection of sediment for porewater analysis and core incubation experiments. U.S. Geological Survey WAWSC staff Greg B. Justin and Peter H. Laird and former U.S. Geological Survey Columbia River Research Laboratory staff Gray L. Rust operated boats for collection of marine water and sediment. Former U.S. Geological Survey WAWSC staff Kelly L. Scholting oversaw the processing of samples for TSS analyses, weighed all filters, and calculated the final TSS results with laboratory assistance from Cg Laird and Karen L. Payne. Current and former U.S. Geological Survey WAWSC staff Brietta Carter, James R. Foreman, Morgan E. Keys, Karen L. Payne, Richard W. Sheibley and Kelly L. Scholting assisted in the laboratory processing of marine surface, porewater and incubation cores. Wisconsin Mercury Research Laboratory (Middleton, WI) staff Jacob, M. Ogorek, Thomas G. Sabin and Charles D. Thompson performed the analysis of FTHg, FMHg and PTHg. Jim Dobbs of the NWQL analyzed FTHg in groundwater and industrial sources. The data on marine currents in Sinclair Inlet analyzed in this report was collected in 1994 by a consortium of researchers from the U.S. Geological Survey. Former USGS employee Jeff Gartner provided access to the data archive for the 1994 ADCP deployments.

References Cited

Albertson, S.L., Newton, J., Eisner, L., Janzen, C., and Bell, S., 1995, 1992 Sinclair and Dyes Inlet seasonal monitoring report, Washington State Department of Ecology Report 95-345, Olympia, Wash., 91 p., accessed August 2, 2012, at https://fortress.wa.gov/ecy/publications/summarypages/95345.html.

Babiarz, C.L., Hurley, J.P., Hoffman, S.R., Andren, A.W., Shafer, M.M., and Armstrong, D.E., 2001, Partitioning of total mercury and methylmercury to the colloidal phase in freshwater: Environmental Science and Technology, v. 35, p. 4,773–4,782.

Balogh, S.J., Nollet, Y.H., Swain, E.B., and others 2004, Redox chemistry in Minnesota streams during episodes of increased methylmercury discharge: Environmental Science and Technology, v. 38, no.19, p. 4,921–4,927.

Battelle Duxbury, 1998, CW/QAPP for Water column monitoring, 1998–2000: MWRA Contract No. S274, Duxbury, MA, 76 p.

Brandenberger, J.M., Louchouarn, P., Kuo, L-J., Crecelius, E.A., V. Cullinan, Gill, G.A., Garland, C., Williamson, J., and Dhammapala. R., 2010, Control of toxic chemicals in Puget Sound, phase 3—Study of atmospheric deposition of air toxics to the surface of Puget Sound: Washington State Department of Ecology Report 10-02-012, Olympia, Wash., 90 p.

Brandenberger, J.M., May, C., Cullinan V., Johnston, R.K., Leisle, D.E., Beckwith, B., Sherrell, G., Metallo, D., and Pingree, R., 2007, 2003–2005 contaminant concentrations in storm water from Sinclair/Dyes Inlet watershed a subbasin of Puget Sound, WA, USA, in Georgia Basin Puget Sound Research Conference, 2007, Proceedings: Puget Sound Action Team and Environment Canada, 9 p., accessed August 2, 2012, at http://depts.washington.edu/uwconf/2007psgb/2007proceedings/oral_abstracts html#9F

Charles, M.J., and Hites, R.A., 1987, Sediments as archives of environmental pollution trends, in R.A. Hites, and Eisenreich, S.J., eds., Sources and fate of aquatic pollutants: Washington D.C., American Chemical Society, p. 365–389.

Choe, K-Y, Gill, G.A., Lehman, R.D., Han, S., Heim, W.A., and Coale, K.H., 2004, Sediment-water exchange of total mercury and nonmethyl mercury in the San Francisco Bay Delta: Limnology and Oceanography, v. 49, no. 5, p 1,512–1,572.

City of Bremerton, 2012, CSO Reduction Program Overview, completed in 2009: City of Bremerton web site, accessed September 13, 2012, at http://www.cityofbremerton.com/content/cso_programoverview.html.

Cokelet, E.D., and R.J. Stewart, 1985, The exchange of water in fjords—The efflux/reflux theory of advective reaches separated by mixing zones: Journal of Geophysical Research, v. 90, no. D3, p. 7,287–7,306.

Cullinan, V.I., May, C.W., Brandenberger, J.M., Judd, C., and Johnston, R.K., 2007, Development of an empirical water quality model for stormwater based on watershed land-use in Puget Sound, in Georgia Basin Puget Sound Research Conference, 2007, Proceedings: Puget Sound Action Team and Environment Canada, access August 2, 2012 at http://depts.washington.edu/uwconf/2007psgb/2007proceedings/oral html.

ENVironmental inVESTment, 2006, Puget Sound naval shipyard and intermediate maintenance facility project ENVVEST community update CD—Study plans, reports, data and supporting information: Prepared by Space and Naval Warfare Systems Center San Diego, Marine Environmental Support Office–NW, Bremerton, Wash., Washington State Department of Ecology Publication Number 06-10-054, August 2006, accessed August 2, 2012, at http://environ.spawar.navy.mil/Projects/ENVVEST/ENVVEST2006/index html.

Evans-Hamilton, Inc. and D.R. Systems, Inc., 1987, Puget Sound Environmental Atlas: Prepared for U.S. Environmental Protection Agency, Puget Sound Water Quality Authority, and U.S. Army Corps of Engineers, U.S. Environmental Protection Agency Report No. EPA-910-9-87-171, 2 v.

Gartner, J.W., Prych, E.A., Tate, G.B., Cacchione, D.A., Cheng, R.T, Bidlake, W.R., and Ferreria, J.T., 1998, Water Velocities and the potential for the movement of bed sediments in Sinclair Inlet of Puget Sound, Washington: U.S. Geological Survey Open File Report 98-572, 140 p.

Gbondo-Tugbawa, S.S., McAlear, J.A., Driscoll, C.T., and Sharpe, C.W., 2010, Total and methyl mercury transformations and mass loadings within a wastewater treatment plant and the impact of the effluent discharge to an alkaline hypereutrophic lake: Water Research, v. 44, no. 9, p. 2,863–2,875.

Grassi, S., Netti, R., 2000, Sea water intrusion and mercury pollution of some coastal aquifers in the province of Grosseto (Southern Tuscany - Italy): Journal of Hydrology, v. 237, p. 98–211.

Helsel, D.R and Hirsch, R.M. 1992, Statistical methods in water resources: Amsterdam, Studies in Environmental Science 49, 522 p.

Huffman, R.L., Wagner, R.J., Toft, J., Cordell, J., DeWild, J.F., Dinicola, R.S., Aiken, G.R., Krabbenhoft, D.P., Marvin-DiPasquale, M., Steward, A.R., Moran, P.W., and Paulson, A.J., 2012, Mercury species and other selected constituent concentrations in water, sediment, and biota of Sinclair Inlet, Kitsap County, Washington, 2007–10: U.S. Geological Survey Data Series 658, 64 p. (Also available at http://pubs.usgs.gov/ds/658/.)

Johnston, R.K., Leisle, D.E., Brandenberger, J.M., Steinert, S.A., Salazar, M., and Salazar, S., 2007, Contaminant residues in demersal fish, invertebrates, and deployed mussels in selected areas of the Puget Sound, WA—Proceedings of the 2007 Georgia Basin Puget Sound Research Conference, Puget Sound Action Team and Environment Canada, 10 p., accessed August 2, 2012, at http://depts.washington.edu/uwconf/2007psgb/2007proceedings/oral_abstracts.html#13E.

Johnston, R.K., Wang, P.F., Loy, E.C., Blake, A.C., Richter, K.E., Brand, M.C, Skahill, B.E., May, C.W., Cullinan, V., Choi, W., Whitney, V.S., Leisle, D.E., and Beckwith, B., 2009, An Integrated Watershed and Receiving Water Model for Fecal Coliform Fate and Transport in Sinclair and Dyes Inlets, Puget Sound, WA: Space and Naval Warfare Systems Center, Technical Report 1977, Dec. 2, 2009, 151 p., accessed August 2, 2012, at http://environ.spawar navy.mil/Publications/pubs2 html.

Johnston, R.K., Rosen G.H., Brandenberger, J.M., Mollerstuen, E.W., Young, J.M., and Beckwith, B., 2011, Monitoring water, sediment, and biota to assess protection of beneficial uses for Sinclair Inlet—Proceedings of Salish Sea Ecosystems Conference 2011, Vancouver, British Columbia, Canada, accessed August 2, 2012, at http://www.verney.ca/assets/O2AProceedings_Johnston.pdf.

Katz, C.N., Noble, P.L., Chadwick, D.B., Davidson, B., and Gauthier, R.D, 2004, Sinclair Inlet water quality assessment: San Diego, Calif., Water Quality Surveys conducted September 1997, March 1998, and July 1998: Space and Naval Warfare Systems Center, accessed August 2, 2012, at http://environ.spawar.navy.mil/Projects/ENVVEST/ENVVEST2006/Reports/ECOS_Survey_Rpt.htm.

Khangaonkar, T., Yang, Z., Taeyun, K., and Roberts, M., 2011, Tidally averaged circulation in Puget Sound sub-basins—Comparison of historical data, analytical model, and numerical model: Estuarine, Coastal and Shelf Science, v. 93, no. 4, p. 305–319.

King County Department of Natural Resources and Parks, 2001, Water quality status report for marine waters, 1999 and 2000: Seattle, Wash., King County Department of Natural Resources and Parks, 558 p.

Knudsen, M., 1900, Ein hydrographischer Lehrsatz: Annalen der Hydrographie und Maritimen Meteorologie, v. 28, p. 316–320.

Lewis, M.E., and Brigham, M.E., 2004, Low-Level Mercury, National Field Manual for the Collection of Water-Quality Data: U.S. Geological Survey Techniques of Water-Resources Investigations, book 9, sec. 6.4B, chap. A5, 26 p.

Luengen, A.C., and Flegal, A.R, 2009, Role of phytoplankton in mercury cycling in the San Francisco Bay estuary: Limnology and Oceanography, v. 54, no. 1, p. 23–40.

Malins, D.C., McCain, B.B., Brown, D.W., Sparks, A.K., and Hodgins, H.,1982, Chemical contaminants and abnormalities in fish and invertebrates from Puget Sound: Boulder, Colo., National Oceanic and Atmospheric Administration Technical Memorandum, OMPA-19, 168 p.

Marvin-DiPasquale, M., and Agee, J.L., 2003, Microbial mercury cycling in sediment of San Francisco, Bay-Delta: Estuaries, v. 26, no. 6, p. 1,517–1,528.

May, C.W., Cullinan, V.I., Woodruff, D., Evans, N., O'Rourke, L., Miller, L., Johnston, R.K., Wang, P.F., Halkola, H., Richter, K.E., B. Davidson, D., Whitney, V., and Right, J., 2005, An analysis of microbial pollution in the Sinclair-Dyes Inlet watershed: Puget Sound Naval Shipyard & Intermediate Maintenance Facility Project ENVVEST, Washington State Department of Ecology, Publication Number 05-03-042, 428 p., plus appendices, accessed August 2, 2012, http://www.ecy.wa.gov/biblio/0503042.html.

National Atmospheric Deposition Program, 2010, National Atmospheric Deposition Program 2009 Annual Summary: Illinois State Water Survey, University of Illinois at Urbana-Champaign, NADP Data Report 2010-01.

National Oceanic and Atmospheric Administration, 2010, National Climatic Data Center, Satellite and Information Service: National Oceanic and Atmospheric Administration database, accessed on May 2, 2012, at http://www1 ncdc.noaa.gov.

Noble, M.A., Gartner, A.L., Paulson, A.J., Xu, J., and Josberger, E.G., 2006, Transport pathways in the lower reaches of Hood Canal: U.S. Geological Survey Open-File Report 2006-1001, 55 p. (Also available at http://pubs.usgs.gov/of/2006/1001/.)

Paulson, A.J., Feely, R.A., Curl, Jr., H.C., Crecelius, E.A., and Geiselman, T., 1988, The impact of scavenging on trace metal budgets in Puget Sound: Geochimica et Cosmochimica Atca, v. 52, no. 7, p.1,765–1,779.

Paulson, A.J., Keys, M., and Scholting, K, 2010, Mercury in the sediments, waters and biota of Sinclair Inlet, Puget Sound, Washington 1989–2007: U.S. Geological Survey Open-File Report 2009–1285, 220 p.

Prych, E.A., 1997, Numerical simulation of ground-water flow paths and discharge locations at the Puget Sound Naval Shipyard, Bremerton, Washington: U.S. Geological Survey Water-Resource Investigations Report 96-4147, 43 p.

Rantz, S.E., and others, 1982, Measurement and computation of streamflow: U.S. Geological Survey Water-Supply Paper 2175, v. 1. Measurement of Stage and Discharge, 284 p.

Shanley, J.B., Schuster, P.F., Reddy, M.M., Roth, D.A., Tayloer, H.E. and Aiken, G.R., 2002, Mercury on the move during snowmelt in Vermont: Transactions of the American Geophysical Society v. 83: p. 45–48.

Skahill, B.E., and LaHatte, C.,2007, Hydrological simulation program—FORTRAN modeling of the Sinclair–Dyes Inlet watershed for the Puget Sound Naval Shipyard & Intermediate Maintenance Facility Environmental Investment project—FY 2007 Report: Vicksburg, Miss., U.S. Army Engineer Research and Development Center, Waterways Experiment Station, Report to the U.S. Navy Puget Sound Naval Shipyard and Intermediate Maintenance Facility Environmental Division, access August 2, 2012, at http://environ.spawar.navy.mil/Projects/ENVVEST/FC_Model_Report/HSPF_reports.htm, and supporting electronic data accessed June 29, 2011, at the predecessor website to http://environ.spawar navy mil/Projects/ENVVEST/ENVVEST2006/index html.

Tetra Tech, 1990, Puget Sound ambient monitoring program 1989—Marine Sediment Monitoring Program, Olympia, Wash.: Report TD2838 to Washington State Department of Ecology.

URS Consultants, Inc., 1991, Time critical removal action at initial assessment study site 2, Remedial Action Report of November 8, 1991.

U.S. Environmental Protection Agency, 1986, EPA methods 6010, 6020, 7000, 7041, 7060, 7131, 7421, 7470, 7740, and 7841—Test methods for evaluating solid waste—Physical/chemical methods, SW-846, (3d ed.), November 1986, with updates I, II, IIA, IIB and III: Washington D.C., U.S. Environmental Protection Agency Office of Solid Waste and Emergency Response, variously paged.

U.S. Environmental Protection Agency, 2000, EPA superfund record of decision—Puget Sound Naval Shipyard Complex: Seattle, Wash., U.S. Environmental Protection Agency ID WA2170023418 OU 02 BREMERTON, WA, EPA/ROD/R10-00/516, Region X.

U.S. Geological Survey, variously dated, National field manual for the collection of water-quality data: U.S. Geological Survey Techniques of Water-Resources Investigations, book 9, chaps. A1–A9, accessed August 16, 2012, at http://water.usgs.gov/owq/FieldManual/.

U.S. Navy, 1992, Site Inspection Report: Final report prepared by the URS team under contract #N62474-89-0-9295, v. 1, 2,063 p.

U.S. Navy, 1995a, Final remedial investigation report operable unit A—Puget Sound Naval Shipyard, Bremerton, Washington: Final report prepared by the URS Team under contract #N62474-89-0-9295,v. 1, 2,433 p.

U.S. Navy, 1995b, Final remedial investigation report operable unit NSC fleet and industrial supply center, Bremerton, Washington: Final report prepared by the URS Team under contract #N62474-89-0-9295, v. 1, 1,956 p.

U.S. Navy, 1999, Sediment characterization at Puget Sound Naval Shipyard, Bremerton, Washington—Phase 1: Prepared by Science Applications International Corporation, 110 p.

U.S. Navy, 2002, Final remedial investigation report operable unit B, Bremerton naval complex, Bremerton, Washington: Final report prepared by the URS Grier under contract #N62474-89-0-9295, v. 1, 1,402 p.

U.S. Navy, 2006a, 2003 marine monitoring report OU B marine, Bremerton naval complex Bremerton, Washington: Final report prepared by the URS Group, Inc. under contract No. N44255-02-D-2008, 295 p.

U.S. Navy, 2006b, 2005 marine monitoring report OU B marine, Bremerton naval complex Bremerton, Washington: Final Report prepared by the URS Group., Inc. under contract No. N44255-02-D-2008, 272 p.

U.S. Navy, 2007a, Final long-term monitoring trend analysis report for OU A, OU NSC, OU B T, PMP, and OU C, Bremerton naval complex, Bremerton, Washington: Prepared for NAVFAC NW by SES-TECH under Contract No. N44255-05-D-5101, Task Order 01, February 2007.

U.S. Navy, 2007b, Sediment transport study and natural recovery model report, BNC Bremerton, Washington: Seattle, Wash., URS Group., Inc., 224 p.

U.S. Navy, 2008a, Second five-year review, Bremerton naval complex, Bremerton, Washington: Silverdale, Wash., U.S. Navy Naval Facilities Engineering Command Northwest, 258 p.

U.S. Navy, 2008b, 2007 marine monitoring report OU B marine, Bremerton naval complex Bremerton, Washington: Final report prepared by the URS Group, Inc. under contract No. N44255-05-D-5100.

U.S. Navy, 2009, Long-term monitoring 2007/2008 trend analysis report for OU A, OU NSC, OU B T, PMP, and OU C, Bremerton naval complex, Bremerton, Washington: Final report prepared by the URS Group, Inc. under contract No. 44255-05-D-5101, Task order 49, April 2009. 224 p.

U.S. Navy, 2012 Final 2010 marine monitoring Report, Bremerton naval complex: Final report prepared by the URS Group, Inc. under contract No. N44255-09-D-4001 April 24, 2012.

Wang, P.F., Johnston, R.K., Halkola, H., Richter, R.E., and Davidson, B., 2005, A modeling study of combined sewer overflows in the Port Washington Narrows and fecal coliform transport in Sinclair and Dyes Inlets, Washington: San Diego, Calif., Space and Naval Warfare Systems Center, Puget Sound Naval Shipyard and Intermediate Maintenance Facility Project ENVVEST, Final Report of June 18, 2005, accessed August 2, 2012, at http://environ. spawar.navy.mil/Projects/ENVVEST/ENVVEST2006/ Reports/Animations/DrougueStudy/DrogueStudyOverview. htm.

Wang, P.F., and Richter, K.E., 1999, A hydrodynamic modeling study using CH3D for Sinclair Inlet: San Diego, Calif., Space and Naval Warfare Systems Center, accessed August 2, 2012, at http://environ.spawar.navy.mil/Projects/ ENVVEST/FC_Model_Report/CH3D_reports htm.

Washington State Department of Ecology and King County, 2011, Control of toxic chemicals in Puget Sound— Assessment of selected toxic chemicals in the Puget Sound Basin, 2007–2011: Olympia, Wash., Washington State Department of Ecology, and Seattle, Wash., King County Department of Natural Resources, Ecology Publication No. 11-03-055, 295 p., accessed August 2, 2012, at http://www. ecy.wa.gov/biblio/1103055.html.

Washington State Department of Ecology, 2012a, Long-term marine water quality data: Washington State Department of Ecology database, accessed August 2, 2012, at http://www. ecy.wa.gov/apps/eap/marinewq/mwdataset.asp.

Washington State Department of Ecology, 2012b, Sinclair/ Dyes Inlets Water Quality Improvement Project: Washington State Department of Ecology web site, accessed September 12, 2012, at http://www.ecy.wa.gov/programs/ wq/tmdl/sinclair-dyes_inlets/.

Appendix A. Data From U.S. Navy and Environmental Investment Projects

Site 2 Soils Data

Concentrations of THg (table A1) in the soils from the monitoring wells and boreholes were measured by Inductively Coupled Plasma (ICP) (Method 6010, U.S. Environmental Protection Agency, 1986). Samples were processed within the required holding times. The initial and continuing calibration verification criteria for THg were met. The contract required detection limit (CRDL) standards, interference check samples (ISC), laboratory control samples (LCS), and the matrix spike samples (MS) all fell within the acceptable recovery range and duplicate sample RPD calculations were within control limits. Continuing calibration blanks were not run. THg was not detected in the field blanks or preparation blanks. In the field, replicate samples were collected at PS02-MW01 and PS02-MW04W (fig. 5), with a relative percent difference of 25 and 18 percent respectively (U.S. Navy, 1992).

Analysis of WTHg Concentrations in Freshwater by the ENVironmental inVESTment project

Unfiltered surface-water samples (streams and storm drains) and wastewater effluent were collected during the ENVVEST project from 2002 to 2005 and analyzed at Battelle Marine Sciences Laboratory in Sequim, Washington, using cold vapor atomic fluorescence spectrometry (CVAFS) in accordance with Battelle SOP MSL-I-013, Total Mercury in Aqueous Samples by CVAFS, following U.S. Environmental Protection Agency method 1631, revision E (ENVironmental inVESTment Project, 2006). The collection of quality control samples included equipment, method, and laboratory controls blanks to indicate bias and contamination; field and laboratory replicates to indicate precision; and standard reference samples to indicate accuracy.

The equipment blanks apparently were processed in the field, thus providing a measure of bias and contamination from both the equipment and the field sampling and processing procedures. Method blanks and blanks for recovery of laboratory controls were processed in the laboratory to indicate bias and contamination that may exist in the analytical procedures. The quality of equipment blanks processed in 2002 was excellent and the quality of method blanks processed from 2002 through 2005 was generally quite good. In 2002, six WTHg concentrations associated with equipment blanks were less than 0.5 ng/L, while the WTHg concentration for the 2005 equipment blank sample was 0.277 ng/L (table A2). In 2003, 2004, and 2005, WTHg was not detected in 14 of the 19 method blanks at reporting level that ranged from 0.1 to 0.17 ng/L. The maximum WTHg concentration in the five method blank samples was 0.221 ng/L. The median blank concentration for recovery of laboratory control samples (LCS) from 2002 through 2005 was 0.293 ng/L. Project reporting limits from 2002 through 2005 ranged from a minimum 0.38 ng/L in 2004 to 0.541 in 2005.

Table A1. Concentrations of total mercury in soils from boreholes and the construction of monitoring wells in Site 2 during the Initial Assessment Study in 1990, Kitsap County, Washington.

[Data source: U.S. Navy (1992). Locations are shown in figure 5. **Location name:** PS02, site 2 of Initial Assessment Study; MW, monitoring well; H, borehole. **Abbreviations:** ft, foot; mg/kg, milligram per kilogram]

Location name	Date collected	Depth range (feet)	Total mercury (mg/kg)
PS02-MW01	06-20-90	2–4	24
	06-20-90	2–4	31
	06-21-90	18–20	29
	06-21-90	23–24	16
PS02-MW02	06-22-90	18–20	26
	06-22-90	22–24	20
PS02-MW03	06-25-90	2–4	24
	06-25-90	12–14	21
	06-25-90	20–22	29
PS02-MW04W	06-26-90	2–4	31
	06-26-90	8–10	24
	06-26-90	14–16	31
	06-26-90	14–16	26
PS02-MW05	06-21-90	2–4	22
	06-21-90	32–34	25
	06-21-90	37–38	21
H-101	06-14-90	26–28	25
	06-14-90	30–32	11
H-102	06-12-90	2–4	18
	06-12-90	14–15	18
	06-12-90	16–18	16
H-104	06-19-90	2–4	14
	06-19-90	16–17	16
	06-19-90	19–20	7.9
H-105	06-14-90	19–20	15
	06-14-90	24–25	12
H-106	06-19-90	16–18	6.6
	06-19-90	16–18	15
	06-19-90	20–22	16
H-107	06-19-90	13–14	18
	06-19-90	15–16	17
H-108	06-15-90	14–18	15
	06-15-90	19–20	13
H-110	06-20-90	11–12	13
	06-20-90	15–16	11
H-111	06-15-90	14–16	14
	06-15-90	17–18	13
H-113	06-20-90	12–13	13
	06-20-90	14–16	12
H-115	06-26-90	2–14	29
	06-26-90	11–12	20
	06-26-90	13–14	17
Minimum			6.6
Maximum			31
Median			17.5

Thirteen laboratory replicates were processed to determine precision of laboratory procedures, and results were within the expectations of the project data-quality objectives. The results provide an indicator of precision and were within the project data-quality objective acceptance criteria of 30 percent for replicate precision. Relative percent differences of WTHg concentrations in replicate sample ranged from 0 to 28.1 percent and the median relative percent difference was 8.8 percent (table A3).

Standard reference materials (SRM) were analyzed at the laboratory to determine accuracy of analytical procedures. Percent differences between the analytical results and the SRM certified value ranged from 0.2 to 10.3 percent (table A4). The median percent difference was 2.9 percent. These percentages are well within the project data quality objectives acceptance criteria of 20 percent.

Matrix spike samples were analyzed and the percent recovery serves as a measure of accuracy of the analytical procedures. Additionally, the matrix spikes were analyzed in duplicate pairs, thus providing another measure of precision for the pair of matrix spike recoveries. Percent recoveries of WTHg in matrix spike samples ranged from 79.3 to 116.3 percent (median: 102 percent) and were within the project data quality objectives acceptance criteria of 70–130 percent (table A5). Precision of the matrix spike duplicates ranged from 1.7 to 19.4 percent (median: 2.0 percent) and were within the project data quality objectives acceptance criteria of 30 percent.

Concentrations of THg and ancillary date in unfiltered creek water are presented in table A6. The statistics for the regression of WTHg versus TSS concentrations are given in table A7 and shown in figure A1. Concentrations of THg from samples collected from stormwater drains discharging outside of the BNC (table A8), wastewater treatment plant effluents (table A9), and BNC stormwater drains (table A10) are presented along with ancillary data. The regression of WTHg versus TSS are shown in figure A2.

Groundwater Data

Prior to 2008, all FTHg and WTHg concentrations in BNC groundwater sampled for Initial Assessment Study (1990), Site Investigation (1990–91), Remedial Investigation/ Feasibility Studies (1993–95), Synoptic groundwater Monitoring (1998–2002), and LTMP (2004–07) were analyzed by older, relatively insensitive cold-vapor atomic absorption spectrometric (CV-AAS) methods with minimum detection limits (MDL) on the order of 100–200 ng/L. FTHg was not detected in any of the 235 samples collected before 2004. Detection of WTHg concentrations above the MDL likely were caused by the presence of high concentrations of solids in the pumped groundwater for WTHg analysis as documented by high turbidity and high concentrations of TSS and total aluminum. Between 2004 and 2007, the MDL of the analysis of FTHg and WTHg in OU B Terrestrial groundwater was also between 100 and 200 ng/L (U.S. Navy, 2007a). One well (LTMP-3) contained FTHg concentrations from slightly higher than the MDL to about 5 times the MDL. Groundwater from two wells (LTMP-3 and -5) contained WTHg significantly higher than the MDL, whereas two other wells (LTMP-1 and 410R) contained WTHg slightly higher than the MDL.

Beginning in 2008, groundwater samples from the LTMP were analyzed by more sensitive cold-vapor atomic absorption fluorescence spectrometric methods (CV-AFS) with a MDL of 1 ng/L (table A11, U.S. Navy, 2009). THg was not detected in the trip blanks or the rinsate blanks. For 2008 and 2009, all quality control samples were within the control limits. Results of samples from three wells with the highest concentrations greater than 10 ng/L (in order of LTMP-3, LTMP-1, LTMP-5) confirm that results of earlier studies using the CV-AAS method was best limited to screening purposes. The quantitative use of earlier data generated by CV-AAS method should be conducted with extreme caution.

During the autumn 2008, sampling a field duplicate of water from PS09-MW01B (RPD of 14) was collected two weeks after the first sample. The WTHg concentration of water from PS07-MW04 was lower than the detection limit. Samples from wells OUB-MW15 and PS07-MW03 yielded estimated positive values. All other 2008 WTHg samples yielded acceptable values greater than the 1ng/L reporting limit.

Apart from the non-detectable concentrations measured for both field replicates from well PS07-MW04 sampled during spring 2009, all samples had acceptable values greater than the reporting limit. One well sampled in March 2009 had duplicate WTHg concentrations lower than the MDL. All WTHg results from wells LTMP-3 and LTMP-5 are presented in table A12.

Table A2.   Concentrations of total mercury in equipment, method, and laboratory-control blanks, by the ENVironmental inVESTment Project, 2002–05.

[Data Source: ENVironmental inVESTment Project (2006). **Laboratory qualifier:** U, analyte not detected at or above the method detection limit; J, analyte detected above the method detection limit, but below the reporting limit. **Abbreviations:** ng/L, nanogram per liter; –, not applicable]

Sampling event	Date of collection	Sample identifier	Mercury concentration (ng/L)	Laboratory qualifier
Equipment blanks				
2002-Equipment blanks	09-16-02	EB-SW4-1	0.5	U
	09-17-02	EB-M6-1	0.5	U
	09-18-02	EB-PA-1	0.5	U
	09-19-02	EB-LMK136-1	0.5	U
	09-20-02	EB-SWPSNS2-1	0.5	U
2005-Storm season	12-03-04	BST12-RB	0.277	J
Method blanks				
2002-Baseflow	10-2002	BLANK 100402	0.500	U
	10-2002	BLANK 100702	0.500	U
	10-2002	BLANK 100802	0.500	U
	10-2002	BLANK 100902	0.717	–
2003-1,3	01-08-03	Method blank (5)	0.115	J
	02-05-03	Method blank (6)	0.122	J
2003-2	01-23-03	Method blank (3)	0.108	U
2003-4	02-10-03	Mean method blank	0.088	U
2003-5,6,7	03-21-03	Method blank (3)	0.100	U
2004-1	05-06-04	Method blank (1)	0.173	J
2004-2	06-22-04	Method blank -1	0.191	J
	06-23-04	Method blank -2	0.221	J
2004-3	12-02-04	Method blank -1	0.116	U
2004-Marine boundary event 1	07-13-04	Mean method blank	0.120	U
2004-Marine boundary event 2	12-03-04	Method blank -1	0.120	U
	12-03-04	Method blank -2	0.120	U
	12-03-04	Method blank -3	0.120	U
2005-Marine 1	02-20-05	Method blank	0.170	U
2005-Marine 2	03-11-05	Mean method blank	0.170	U
2005-Marine 3	03-30-05	Mean method blank	0.170	U
2005-Marine 4	04-07-05	Mean method blank	0.170	U
2005-Marine 5	06-29-05	Mean method blank	0.170	U
2005-Marine 6	09-26-05	Mean method blank	0.170	U
Laboratory Control blanks				
2003-1,3	01-08-03	BLANK010703	0.270	J
	02-05-03	BLANK020403	0.271	J
2003-2	01-23-03	BLANK012203	0.279	J
2003-4	02-10-03	BLANK020703	0.249	–
2003-5,6,7	03-21-03	BLANK032003	0.211	J
2004-1	05-06-04	BLANK050504	0.361	J
2004-2	06-22-04	BLANK062104	0.300	J
	06-23-04	BLANK062204	0.302	J
2004-3	12-02-04	BLANK120104	0.293	J
2004-Marine boundary event 1	07-13-04	BLANK071204	0.263	J
2004-Marine boundary event 2	12-03-04	BLANK120204	0.301	J
2005-Marine 1	02-20-05	BLANK022105	0.205	J
2005-Marine 2	03-11-05	BLANK031005	0.308	J
2005-Marine 3	03-30-05	BLANK032905	0.262	J
2005-Marine 4	04-07-05	BLANK040605	0.398	J
2005-Marine 5	06-29-05	BLANK062805	0.363	J
2005-Marine 6	09-26-05	BLANK092305	0.400	J
		Median	0.293	

Table A3.   Relative percent differences between replicate analyses of total mercury in unfiltered water, ENVironmental inVESTvestment Project, 2002–05.

[Samples analyzed at Battelle Marine Science Laboratories, Sequim, Washington. Data source: ENVironmental inVESTment Project (2006). **Abbreviations:** µg/L, microgram per liter; r or R, replicate; DUP, duplicate; – not applicable]

Sampling event	Sample date	Sample identifier	Total mercury concentration (µg/L)	Mean replicate concentration (µg/L)	Relative percent difference
2002	09-18-02	1783-297 r1	0.00214	–	–
	09-18-02	1783-297 r2	0.00241	0.00227	12.0
	09-16-02	1783-179 r1	0.00228	–	–
	09-16-02	1783-179 r2	0.00239	0.00234	4.5
	09-18-02	1783-274 r1	0.00079	–	–
	09-18-02	1783-274 r2	0.00105	0.00092	28.1
	09-17-02	1783-240 r1	0.00678	–	–
	09-17-02	1783-240 r2	0.00592	0.00635	13.6
2003-1, 3	12-15-02	1937-29 r1	0.00923	–	–
	12-15-02	1937-29 r2	0.00923	0.00923	0.0
2003-2	01-11-03	1937-80 r1	0.00753	–	–
	01-11-03	1937-80 r2	0.00625	0.00689	18.6
2003-4	01-29–30-03	1937-358 r1	0.00546	–	–
	01-29–30-03	1937-358 r2	0.00544	0.00545	0.3
2003-5, 6, 7	02-16-03	1937-445 r1	0.00233	–	–
	02-16-03	1937-445 r2	0.00242	0.00237	3.8
2004-1	05-06-04	2140-59 R1	0.00430	–	–
	05-06-04	2140-59 R2	0.00427	0.00428	0.8
2004-2	06-22-04	2140*99 R1	0.02575	–	–
	06-22-04	2140*99 R2	0.02359	0.02467	8.8
	06-23-04	2140*202 R1	0.02773	–	–
	06-23-04	2140*202 R2	0.03038	0.02906	9.1
2004-3	12-02-04	2140-244	0.01276	–	–
	12-02-04	2140-244DUP	0.01100	0.01188	14.8
2005	03-11-05	2318*176 R1	0.00754	–	–
	03-11-05	2318*176 R2	0.00728	0.00741	3.4
				Median	8.8

Table A4.   Percent differences of concentrations of standard reference materials for analysis of total mercury in water, ENVironmental inVESTvestment Project, 2002–05.

[Data Source: ENVironmental inVESTment Project (2006). Standard Reference Material 1641d analyzed at Battelle Marine Science Laboratories, Sequim, Washington. **Abbreviations:** ng/L, nanogram per liter; –, not available or applicable]

	Standard Reference Material 1641d mercury in natural water					
Sampling event	Sample identifier	Analysis date		Total mercury concentration (ng/L)	Percent recovery	Percent difference
2002-Baseflow	1641d100402	–		1,495,663	94.1	5.9
	1641d100702	–		1,609,726	101.2	1.2
	1641d100802	–		1,595,292	100.3	0.3
	1641d100902	–		1,592,628	100.2	0.2
			Certified value	1,590,000	–	–
			Range	±40,000	–	–
2003-1,3	1641d 010703	–		1,562,000	98.2	1.8
	1641d 020403	–		1,754,294	110.3	10.3
			Certified value	1,590,000	–	–
			Range	±40,000	–	–
2003-2	1641d 012203	–		1,636,367	102.9	2.9
			Certified value	1,590,000	–	–
			Range	±40,000	–	–
2003-4	1641d 020703	–		1,439,385	90.5	9.5
			Certified value	1,590,000	–	–
			Range	±40,000	–	–
2003-5, 6, and 7	1641d 032003	–		1,503,042	94.5	5.5
			Certified value	1,590,000	–	–
			Range	±18,000	–	–
2003-Stormwater	1641d 012703	–		1,662,041	104.5	4.5
	1641d 020403	–		1,754,294		10.3
			Certified value	1,590,000	–	–
			Range	±40,000	–	–
2004-1	1641d050404	05-06-04		1,573,000	98.9	1.1
			Certified value	1,590,000	–	–
			Range	±18,000	–	–
2004-2	1641d062104	06-22-04		1,624,000	102.1	2.1
	1641d062204	06-23-04		1,587,000	99.8	0.2
			Certified value	1,590,000	–	–
			Range	±18,000	–	–
2004-3	1641d 120104	12-02-04		1,684,000	105.9	5.9
			Certified value	1,590,000	–	–
			Range	±18,000	–	–
2005	1641d031005	03-11-05		1,521,000	95.7	4.3
	1641d031705	03-17-05		1,569,000	98.7	1.3
			Certified value	1,590,000	–	–
			Range	±18,000	–	–

Table A5.   Percent recovery of matrix spikes and relative percent differences of matrix spike for analysis of total mercury in water measured by the Environmental iInVESTvestment Project, 2002–05.

[Data Source: ENVironmental inVESTment Project (2006). Samples analyzed at Battelle Marine Science Laboratories, Sequim, Washington. Minor differences in percentages are due to rounding. **Abbreviations:** µg/L, microgram per liter; – not available or applicable]

Sampling event	Sample identifier		Date of collection or analysis	Spiking level (µg/L)	Environ- mental concen- tration (µg/L)	Spike concen- tration (µg/L)	Spike recovered (µg/L)	Percent recovery	Relative percent difference
2002-Baseflow	ALT2-1	1783-193	09-16-02	0.0171	0.00501	0.0220	0.0170	99.4	–
	ALT2-1	1783-193 MSD	09-16-02	0.0180	0.00501	0.0232	0.0182	101.1	1.7
	OC-1	1783-281	09-18-02	0.0117	0.00172	0.0138	0.0121	103.2	–
	OC-1	1783-281 MS	09-18-02	0.0126	0.00172	0.0146	0.0129	102.2	1.0
	SWPSNS2-1	1783-309	09-18-02	0.0140	0.0211	0.0358	0.0147	105.0	–
	SWPSNS2-1	1783-309 MSD	09-18-02	0.0143	0.0211	0.0345	0.0134	93.7	11.4
	M7-1	1783-234	09-17-02	0.0115	0.000820	0.0127	0.0119	103.3	–
	M7-1	1783-234 MS	09-17-02	0.0113	0.000820	0.0125	0.0117	103.4	0.1
	M1-1	1783-218 R1	09-17-02	0.0111	0.000862	0.0122	0.0113	102.1	–
	M1-1	1783-218 MS	09-17-02	0.0107	0.000862	0.0120	0.0111	104.1	1.9
2003-1, 3	CH-1	1937-19	01-08-03	0.0264	0.00444	0.0254	0.0210	79.4	–
	CH-1	1937-19	01-08-03	0.0269	0.00444	0.0262	0.0218	81.0	1.9
2003-2	SC-2	1937-90	01-11-03	0.0122	0.00616	0.0190	0.0128	105.0	–
	SC-2	1937-90 MS	01-11-03	0.0105	0.00616	0.0168	0.0106	101.4	3.5
2003-4	BL-4	1937-364	02-10-03	0.0171	0.00239	0.0202	0.0178	104.2	–
	BL-4	1937-364 MS	02-10-03	0.0153	0.00239	0.0175	0.0151	98.3	5.8
2003-5, 6, and 7	CE-7	1937-538	03-21-03	0.0146	0.0112	0.0255	0.0143	98.2	–
	CE-7	1937-538	03-21-03	0.0130	0.0112	0.0238	0.0126	97.1	1.0
2003-Stormwater	PSNS 115.1	1937-153	01-21-03	0.0181	0.0196	0.0398	0.0202	111.7	–
	PSNS 115.1	1937-153	01-21-03	0.0193	0.0196	0.0420	0.0224	116.3	4.0
	PSNS 115.1	1937-284	01-21-03	0.0210	0.0140	0.0368	0.0229	108.8	–
	PSNS 115.1	1937-284	01-21-03	0.0191	0.0140	0.0354	0.0215	112.3	3.1
2004-1	T1005	LMK 122	05-06-04	0.0108	0.0219	0.0323	0.0105	96.8	–
	T1005	LMK 122	05-06-04	0.0102	0.0219	0.0318	0.0099	97.1	0.4
2004-2	T1011	B-ST/CSO16	06-23-04	0.0186	0.0338	0.0509	0.0170	91.7	–
	T1011	B-ST/CSO16	06-23-04	0.0203	0.0338	0.0532	0.0193	95.4	4.0
	T1014	LMK122	06-23-04	0.0192	0.0498	0.0694	0.0196	102.0	–
	T1014	LMK122	06-23-04	0.0177	0.0498	0.0647	0.0148	83.9	19.4
2004-3	T1018	B-ST28	12-02-04	0.0063	0.0092	0.0154	0.0063	99.493	–
	T1018	B-ST28	12-02-04	0.0059	0.0092	0.0149	0.0058	97.479	2.045
	T1026	2140-248	12-02-04	0.0065	0.0085	0.0154	0.0070	108.009	–
	T1026	2140-248	12-02/04	0.0061	0.0085	0.0149	0.0064	105.675	2.185
2005	G1200 [1]	B-WWTP	03-17-05	0.0180	0.0052	0.0244	0.0193	107.319	–
	G1200	B-WWTP	03-17-05	0.0154	0.0052	0.0214	0.0162	105.619	1.597
							Median	102.1	2.0

[1] Environmental concentration is qualified with a J remark code.

Table A6. Concentrations of total mercury, total aluminum, organic carbon and suspended solids in unfiltered creek waters draining into Sinclair Inlet by the ENVironmental inVESTment Project, Kitsap County, Washington, 2002–05.

[Source of data: ENVironmental inVESTment Project (2006). Stream locations are shown in figure 5. **Abbreviations:** ng/L, nanogram per liter; µg/L, microgram per liter; mg/L, milligram per liter; <, less than; –, not available or not applicable; ND, not detected]

Collection date	Total concentrations				Comment	Field notes
	Total mercury (ng/L)	Total aluminum (µg/L)	Organic carbon (mg/L)	Suspended solids (mg/L)		
Anderson Creek						
03-11-02	24.20	15,000	4.4	327	Wet base flow	
09-18-02	1.20	123	0.6	3	Dry base flow	
01-22-03	8.88	1033	5.0	44	Storm composite	
01-29-03	5.53	908	3.9	16	Storm	
01-30-03	5.45	1,365	–	–	Storm	
02-16-03	3.87	960	–	–	Storm	
01-17-05	12.05	2,050	6	88	Storm	
01-17-05	11.40	2,000	5.8	59	Storm	Independent sample collected at same time (2 samplers)
01-22-05	3.00	215	3.3	5	Storm	
Annapolis Creek						
09-18-02	1.13	55.8	2.3	ND	Dry base flow	
04-19-04	4.11	174.0	4.7	30	Storm	
05-26-04	6.00	667.0	9.6	32	Storm	
10-18-04	5.50	634.0	7.1	29		Average of field duplicates
01-17-05	27.31	3,800	8.0	153	Storm	
01-22-05	4.49	199	6.1	8	Storm	
Blackjack Creek						
03-11-02	9.00	1,750	7.4	67	Wet base flow	
03-12-02	5.27	670	–	13	Wet base flow	
03-13-02	4.90	449	–	10	Wet base flow	
09-16-02	1.39	90.3	3.1	2	Dry base flow	
09-17-02	1.26	68.6	–	6	Dry base flow	
09-18-02	1.06	64.4	–	<2	Dry base flow	
01-22-03	8.68	696	10.9	33	Storm composite	
01-29-03	6.22	593	9.5	13	Storm	
01-30-03	9.01	1,420	–	–	Storm	
02-15-03	3.24	243	5.9	10	Storm	
02-28-05	2.78	179	5.7	7	Storm	
03-19-05	4.85	396	8.4	19	Storm	
03-30-05	3.70	181	10.4	4	–	
Upper Gorst Creek						
03-11-02	11.20	1,810	5.1	51	Wet base flow	
03-12-02	2.99	266	–	10	Wet base flow	
03-13-02	3.69	243	–	7	Wet base flow	
09-16-02	2.94	581	1.9	25	Dry base flow	
09-17-02	2.89	697	–	57	Dry base flow	
09-18-02	4.06	858	–	38	Dry base flow	
09-18-02	0.91	61	1.4	1	Dry base flow	
01-22-03	6.26	569	5.0	50	Storm composite	
01-29-03	3.42	904	3.1	34	Storm	
01-30-03	3.83	953	–	–	Storm	
02-16-03	8.94	1,840	2.4	16	Storm	
01-22-05	2.63	244	2.6	11	Storm	
03-30-05	1.34	87	2.0	2	Dry base flow	

Table A6.   Concentrations of total mercury, total aluminum, organic carbon and suspended solids in unfiltered creek waters draining into Sinclair Inlet by the ENVironmental inVESTment Project, Kitsap County, Washington, 2002–05.—Continued

[Source of data: ENVironmental inVESTment Project (2006). Stream locations are shown in figure 5. **Abbreviations:** ng/L, nanogram per liter; µg/L, microgram per liter; mg/L, milligram per liter; <, less than; –, not available or not applicable; ND, not detected]

Collection date	Total concentrations				Comment	Field notes
	Total mercury (ng/L)	Total aluminum (µg/L)	Organic carbon (mg/L)	Suspended solids (mg/L)		
Lower Gorst Creek at Sam Christopherson						
01-16-05	10.93	1,560	4.9	107	Storm	
01-17-05	9.29	1,390	4.8	63	Storm	
01-22-05	2.66	189	2.9	8	Storm	
Olney Creek						
03-11-02	27.20	8,930	6.8	333	Wet base flow	
09-18-02	1.72	168	1.3	8	Dry base flow	
01-22-03	22.60	2,935	9.4	210	Storm composite	
01-29-03	6.99	1,750	–	57	Storm	
01-30-03	9.55	2,390	5.7	–	Storm	
02-15-03	8.01	1,250	5.6	63	Storm	
02-28-05	7.49	1,190	3.7	59	Storm	
03-19-05	16.96	2,640	5.9	151	Storm	
03-30-05	2.05	228	3	6	Wet base flow	
Summary statistics for all ENVVEST stream data						
Number of samples	53	53	40.0	46		
Average	6.91	1,315	5.10	51		
Standard deviation	6.34	2,364	2.60	75		
Minimum	0.91	55.8	0.6	1		
Maximum	27.31	15,000	10.9	333		

Table A7. Regression of total mercury concentrations against total suspended solids concentrations in unfiltered water samples collected by the Environmental inVESTvestment Project, Kitsap County, Washington, 2002-05.

[Values in **bold** were significant at a probability level of 0.05. **Abbreviations:** ng/L, nanogram per liter; ng/mg, nanogram per milligram; CSO, combined sewer overflow; B, Bremerton; PSNS, Puget Sound Naval Shipyard; PO, Port Orchard; >, greater than; <, less than; –, not applicable]

Location name	Regression of total mercury versus total aluminum	Total suspended solids range (mg/L)	Regression of total mercury versus total suspended solids				
	p value		Number of samples	*p* value	Correlation coefficient	Intercept (ng/L)	Slope (ng/mg)
Watershed							
Blackjack	**<0.001**	<2–67	12	**<0.001**	0.83	2.5	0.12
Anderson	**<0.001**	3–327	7	**<0.001**	0.95	4.5	0.064
Olney	**0.002**	6–333	8	**<0.001**	0.98	2.5	0.082
Annapolis	**<0.001**	1–153	6	**0.001**	0.99	0.92	0.17
Gorst Upper	**<0.001**	1–57	12	0.11	0.49	4.38±1.21	–
Stormwater drains							
CSO16	**>0.001**	<1–75	7	**<0.001**	0.97	0.37	0.42
LMK122	**0.26**	1–92	10	**0.006**	0.79	4.6	0.46
B-ST28	**0.004**	<1–116	7	**0.034**	0.97	2.16	0.19
PO-BLVD	**0.04**	6–149	4	**0.002**	0.99	4.45	0.10
LMK038	**0.008**	2–113	5	**0.005**	0.97	2.54	0.15
B-ST12	0.20	1–253	9	0.06	0.73	–	–
PSNS015	0.23	26–168	5	**0.03**	0.92	-192	7.2
PSNS126	0.24	5–39	6	**0.05**	0.81	7.9	0.46
PSNS124	0.70	8–20	5	0.06	0.86	–	–

Table A8.   Concentrations of total mercury, total aluminum, organic carbon, and suspended solids in unfiltered stormwater draining into Sinclair Inlet measured by the ENVironmental inVESTment Project, Kitsap County, Washington.

[Data Source: ENVironmental inVESTment Project (2006). Stormwater locations are shown in figures 1, 2, and 3. **Abbreviations:** ng/L, nanogram per liter; mg/L, milligram per liter; <, less than; –, not available or not applicable; ND, not detected]

	Total concentrations			
Collection date	**Total mercury (ng/L)**	**Total aluminum (µg/L)**	**Organic carbon (mg/L)**	**Suspended solids (mg/L)**
CSO16 (B-CSO16 Bremerton, Pacific Ave.)				
03-11-02	12.30	1,190	1.1	27
03-13-02	1.90	58	–	ND
03-14-02	1.88	50	–	ND
04-19-04	10.36	566.0	7.1	26
05-26-04	33.8	1,560	10.5	69
10-18-04	27.4	1,710	–	–
02-28-05	32.8	2,390	7.5	75
03-19-05	15.1	1,720	4.3	51
LMK122 (Navy City Metals- Gorst Watershed)				
03-12-02	9.63	2,150	3.9	45
03-13-02	6.23	1,510	–	15
03-14-02	5.03	689	–	9
09-16-02	2.34	69.3	3.7	ND
09-17-02	5.45	241	–	4
09-18-02	1.89	92.2	–	1
04-19-04	21.9	558	6.2	20
05-26-04	49.8	1,050	12.1	48
10-18-04	56.1	802	–	–
01-16-05	41.9	2,640	7.3	92
01-22-05	14.1	642	6	8
ST-28 (Bremerton Callow Ave.)				
03-11-02	6.90	1,057	1.6	24
03-13-02	2.41	120.0	–	2
04-19-04	13.28	1,230	10.4	79
05-26-04	25.27	2,670	16.8	116
10-18-04	9.17	1,168	6.4	31
02-28-05	18.40	2,150	11.7	81
03-19-05	11.06	1,750	5.7	49
PO-BLVD (Port Orchard, Port Orchard Blvd.)				
05-26-04	19.52	3,265	20.5	149
10-18-04	8.46	1,270	6.1	46
01-17-05	12.95	2,230	6.6	87
01-22-05	5.46	419	5.7	6
B-ST12 (Bremerton Trenton Ave.)				
09-16-02	1.17	111	1.9	ND
04-19-04	3.33	186	4.2	5
05-26-04	8.06	957	7.9	34
10-18-04	4.92	850	–	–
03-19-05	5.25	597	3.4	205
03-26-05	5.91	776	2.6	28
03-30-05	44.78	9,930	3	253
03-31-05	9.66	575	2.5	16

Table A8. Concentrations of total mercury, total aluminum, organic carbon, and suspended solids in unfiltered stormwater draining into Sinclair Inlet measured by the ENVironmental inVESTment Project, Kitsap County, Washington.—Continued

[Data Source: ENVironmental inVESTment Project (2006). Stormwater locations are shown in figures 1, 2, and 3. **Abbreviations:** ng/L, nanogram per liter; mg/L, milligram per liter; <, less than; –, not available or not applicable; ND, not detected]

Collection date	Total concentrations			
	Total mercury (ng/L)	Total aluminum (μg/L)	Organic carbon (mg/L)	Suspended solids (mg/L)
LMK038 (Manchester Outfall)				
09-18-02	0.59	36.4	2.5	2
04-19-04	4.28	306	7.6	11
05-26-04	11.25	1,760	11	64
10-18-04	10.80	2,350	–	–
01-17-05	19.23	3,420	6.7	113
01-22-05	6.25	415	6.8	8
Summary statistics for all ENVVEST greater Sinclair Inlet stormwater data				
Number of samples	44	44	33	36
Average	14.05	1,347	6.7	52.8
Standard Deviation	13.67	1,608	4.3	57.4
Minimum	0.59	36	1.1	1.0
Maximum	56.14	9,930	20.5	253.0

Table A9. Concentrations of total mercury, total aluminum, organic carbon, and suspended solids in unfiltered effluents from two wastewater treatment plants ENVironmental inVESTment Project, Kitsap County, Washington, 2004–05.

[Data Source: ENVironmental inVESTment Project (2006). **Abbreviations:** ng/L, nanogram per liter; mg/L, milligram per liter; –, not available]

Collection date	Total concentrations			
	Total mercury (ng/L)	Aluminum (mg/L)	Organic carbon (mg/L)	Suspended solids (mg/L)
City of Bremerton				
05-26-04	17.7	41	15	10
10-19-04	5.62	30	13	–
03-01-05	8.45	27	13	2
03-30-05	34.8	169	11	3
West Sound Utility District				
05-26-04	8.17	98	20	20
01-17-05	32.93	645	25	82
01-22-05	11.58	98	19	14
03-01-05	7.54	50	16	8
03-19-05	6.60	49	19	12
03-26-05	7.15	43	15	9
03-30-05	4.46	19	16	26
04-10-05	39.83	424	17	104

Table A10.   Concentrations of total mercury, total aluminum, organic carbon, and suspended solids in unfiltered stormwater draining from the Bremerton naval complex measured by the ENVironmental inVESTment Project, Kitsap County, Washington, 2002–05.

[Data Source: ENVironmental inVESTment Project (2006). Site identifiers are shown on figure 3. **Abbreviations:** PSNS, Puget Sound Naval Shipyard; ng/L, nanogram per liter; mg/L, milligram per liter; <, less than; –, not available]

Date of sample collection	Site identifier	Total concentrations			
		Total mercury (ng/L)	Total aluminum (mg/L)	Organic carbon (mg/L)	Suspended solids (mg/L)
		PSNS015			
01-21-03	PSNS015	26.09	472	–	–
01-29-03	PSNS015	15.26	127	–	–
04-19-04	PSNS015	181.25	889	5.1	46
05-26-04	PSNS015	1,131.02	2,000	14.5	168
10-18-04	PSNS015	125.00	2,600	5.6	88
02-28-05	PSNS015	65.51	734	7.0	26
03-19-05	PSNS015	116.62	583	6.7	34
Number of samples		7	7	5	5
Average		237	1,058	7.78	72
Standard deviation		398	898	3.84	59
Minimum		15.3	127	5.1	26
Maximum		1,131	2,600	14.5	168
		All other PSNS stormwater drains			
01-21-03	PSNS008	42.80	1,440	–	–
01-29-03	PSNS008	4.92	110	–	–
05-26-04	PSNS008	24.67	977	9.0	45
09-18-02	PSNS011	21.14	80	5.5	1
09-18-02	PSNS052	1.85	90	–	–
01-21-03	PSNS081.1	22.02	311	–	–
01-29-03	PSNS081.1	28.58	206	–	–
01-21-03	PSNS101	5.94	131	–	–
01-29-03	PSNS101	3.80	40	–	–
05-26-04	PSNS101	15.86	145	3.7	32
01-21-03	PSNS115.1	19.56	223	–	–
01-29-03	PSNS115.1	14.68	52	–	–
05-26-04	PSNS115.1	24.80	129	6.4	5
01-21-03	PSNS115.1A	442.91	6,700	–	–
01-29-03	PSNS115.1A	18.55	230	–	–
01-21-03	PSNS124	14.11	262	–	–
01-29-03	PSNS124	19.32	24	–	–
04-19-04	PSNS124	116.31	168	9.3	20
05-26-04	PSNS124	44.01	200	14.4	17
10-18-04	PSNS124	34.75	195	8.9	12
02-28-05	PSNS124	17.23	102	6.1	8
03-19-05	PSNS124	20.46	105	3.8	8
09-18-02	PSNS126	14.77	102	5.3	5
01-21-03	PSNS126	13.04	687	–	–
01-29-03	PSNS126	4.80	222	–	–
04-19-04	PSNS126	21.12	338	14.0	25

Table A10. Concentrations of total mercury, total aluminum, organic carbon, and suspended solids in unfiltered stormwater draining from the Bremerton naval complex measured by the ENVironmental inVESTment Project, Kitsap County, Washington, 2002–05.—Continued

[Data Source: ENVironmental inVESTment Project (2006). Site identifiers are shown on figure 3.
Abbreviations: PSNS, Puget Sound Naval Shipyard; ng/L, nanogram per liter; mg/L, milligram per liter; <, less than; –, not available]

Date of sample collection	Site identifier	Total concentrations			
		Total mercury (ng/L)	Total aluminum (mg/L)	Organic carbon (mg/L)	Suspended solids (mg/L)
All other PSNS stormwater drains—Continued					
05-26-04	PSNS126	27.70	814	20.9	39
10-18-04	PSNS126	12.80	728	7.3	24
02-28-05	PSNS126	11.77	677	9.5	17
03-19-05	PSNS126	27.50	770	17.3	36
05-26-04	PSNS126.1	44.13	159	–	–
Number of samples		31	31	15.0	15
Average		36.64	530	9.4	20
Standard deviation		78.15	1,193	5.1	14
Minimum		1.85	24	3.7	1
Maximum		442.9	6,700	20.9	45

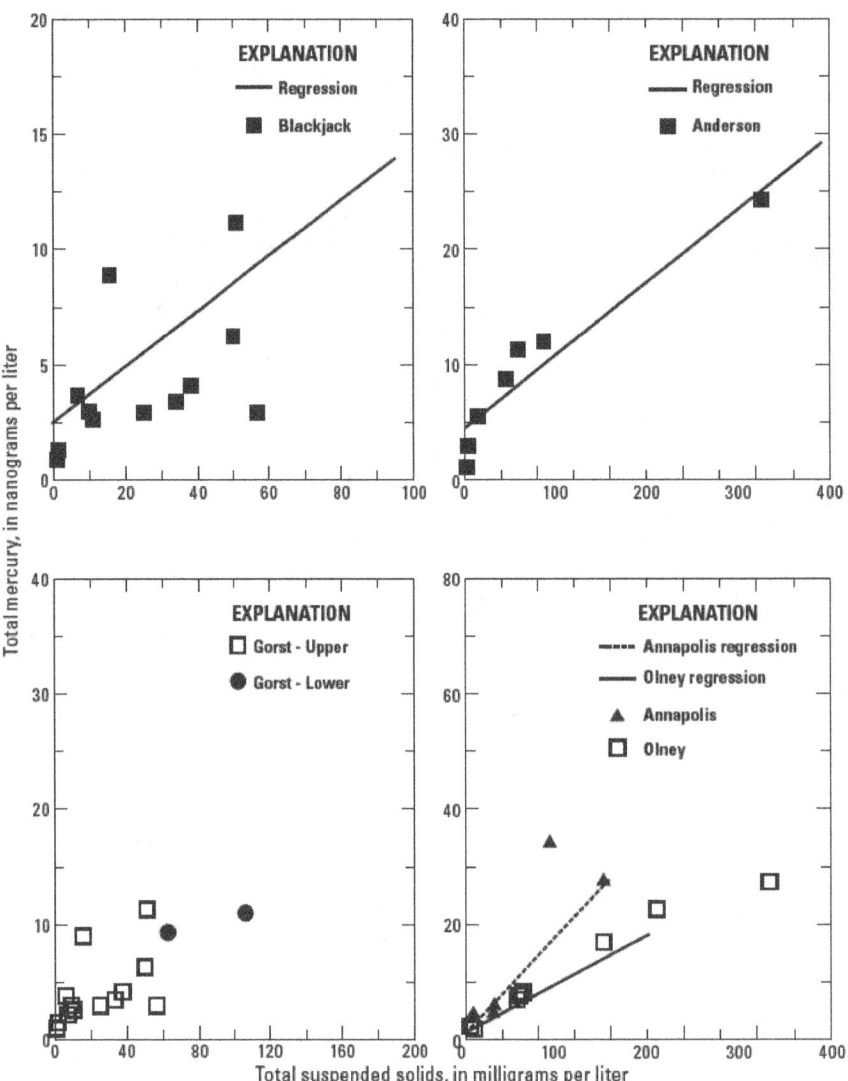

Figure A1.   Total mercury concentrations in unfiltered creek water compared to total suspended solids concentrations with regressions as described in table A7.

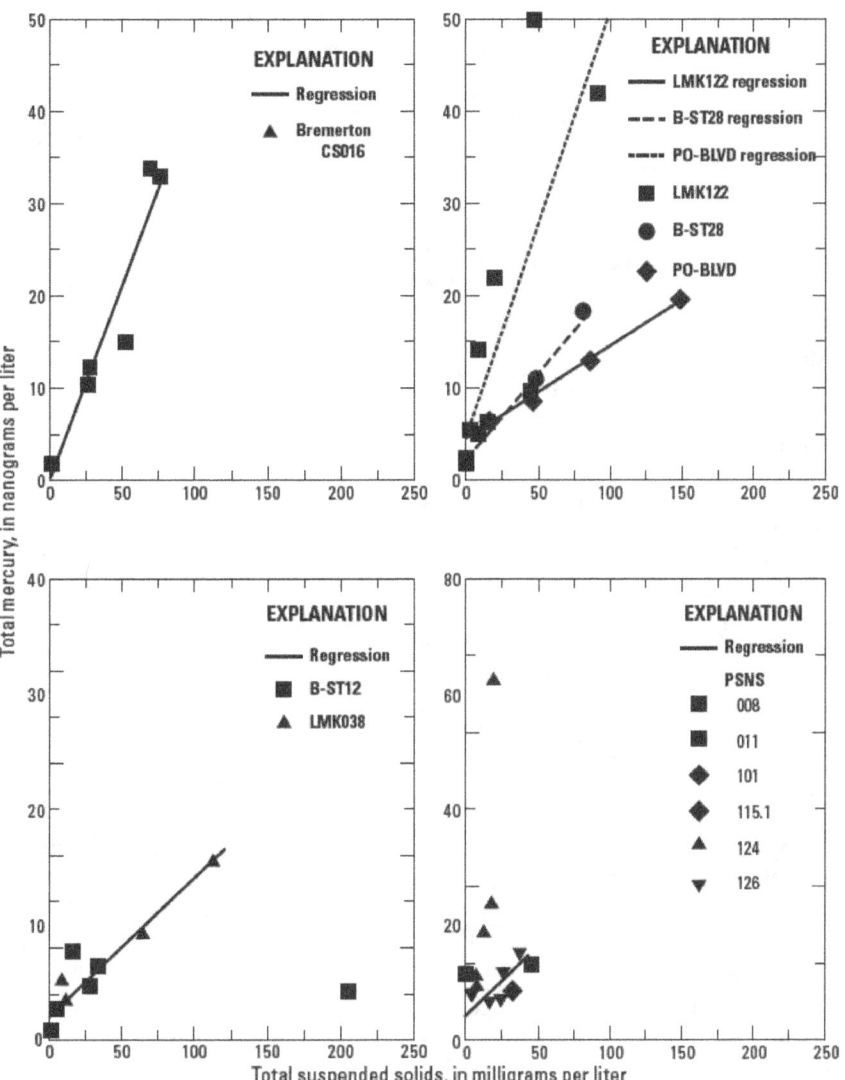

Figure A2. Total mercury concentrations in unfiltered stormwater compared to total suspended solids concentrations with regressions as described in table A7.

Table A11.   Concentrations of total mercury in unfiltered groundwater samples collected from the Operable Unit B-Terrestrial by the Long Term Monitoring Program, 2008–09.

[Data source: Dwight Leisle, U.S. Navy, written commun., 2010. **Abbreviations:** ng/L, nanogram per liter; U, analyte not detected at or above the method detection limit; J, analyte detected above the method detection limit, but below the reporting limit; mg/L, milligram per liter; NTU, nephelometric turbidity units; LTMP, Long-Term Monitoring Program; OUB, operational unit B; PS, Puget Sound; –, not applicable]

Location name	Date sample collected	Total mercury (ng/L)	Final qualifier	Total suspended solids (mg/L)	Turbidity (NTU)
LTMP-1	10-14-08	156	–	–	7
LTMP-2	10-14-08	4.74	–	65	0
LTMP-3	10-14-08	3,680	–	9	22
LTMP-4	10-16-08	3.61	–	13	28
LTMP-5	10-15-08	41.6	–	6	13
OUB-MW15	10-16-08	1.53	J	–	0
OUB-MW18	10-20-08	1.41	–	–	36
PS07-MW03	10-20-08	23.8	J	12	55
PS07-MW04	10-15-08	0.33	U	–	28
PS09-MW01B	10-20-08	26.9	–	37	26
PS09-MW01B	11-05-08	30.9	–	40	30
LTMP-1	04-01-09	56.8	–	–	19
LTMP-2	03-30-09	23.2	–	–	15
LTMP-3	04-02-09	1,190	–	–	30
LTMP-4	03-31-09	2.97	–	–	38
LTMP-5	04-01-09	27.5	–	–	34
OUB-MW15	03-30-09	2.59	–	–	4
OUB-MW18	04-01-09	8.87	–	–	25
PS07-MW03	03-30-09	1.93	–	–	42
PS07-MW04	03-31-09	0.40	U	–	7
PS09-MW01B	03-31-09	10.3	–	–	42
Number of samples		21		7	21
Average		252		26	24
Standard deviation		826		22	15
Minimum		0.33		6	0
Maximum		3,680		65	55

Table A12. Concentrations of total mercury in unfiltered and filtered water and ancillary data from Long-Term Monitoring Program Wells LTMP-3 and LTMP-5 in Operable Unit B-Terrestrial, Bremerton naval complex, Kitsap County, Washington, 2004–07.

[Data source: 2004-07 data, U.S. Navy (2007); 2008-09 data, U.S. Navy (2009); Dwight Leisle, U.S. Navy, written commun., 2010. **Final qualifier:** A, acceptable; J, estimated positive result; U, undetected; B, less than the reporting limit and greater than the instrument detection limit. **Abbreviations:** LTMP, Long-Term Monitoring Program; µg/L, microgram per liter; mg/L, milligram per liter; MP, measuring point; NGVD, National Geodetic Vertical Datum of 1929; MLLW, mean lower low water; –, not analyzed or does not apply]

Date collected	Time collected	Unfiltered mercury (µg/L)	Final qualifier	Filtered mercury (µg/L)	Final qualifier	Unfiltered aluminum (µg/L)	Final qualifier	Total suspended solids (mg/L)	Turbidity	Water level Below MP	Water level NGVD29	Water level MLLW
							LTMP-3 (well OUBT-722)					
08-30-04	–	0.98	A	–	–	5	A	–	–	–	–	–
08-30-04	–	2.4	A	–	–	5	A	–	–	–	–	–
12-01-04	–	2.81	J	–	–	–	–	–	–	–	–	–
12-01-04	–	2.89	J	–	–	–	–	–	–	–	–	–
03-01-05	–	2.48	A	0.97	A	7	A	–	–	–	–	–
05-12-05	–	6.69	A	0.69	A	–	–	–	–	–	–	–
07-14-05	–	–	–	0.21	A	–	–	–	–	–	–	–
07-14-05	–	–	–	0.21	A	–	–	–	9.1	–	-1.02	4.97
10-19-05	–	3.34	A	–	–	31	A	5.09	5.09	–	1.94	7.93
01-19-06	–	2.05	A	–	–	–	–	0	0	–	1.61	7.6
04-12-06	–	3.82	–	–	–	18	–	–	9	–	0.24	6.23
10-10-06	–	4.92	–	–	–	13	–	0.9	0.9	10.26	0.5	6.49
10-10-07	–	3.21	–	–	–	–	–	50	15	11.83	-1.07	4.92
10-14-08	[1]12:21	3.68	–	–	–	–	–	9	22	12.87	-2.11	3.88
04-20-09	[1]13:05	1.19	–	–	–	–	–	–	30	9.78	0.98	6.97
N		13		4		6		5	8			
Average		3.11		0.52		13.2		13	11.4			
Standard deviation		1.5		0.38		10.1		21	10.4			
Minimum		0.98		0.21		5		0	0			
Maximum		6.69		0.97		31		50	30			

Table A12. Concentrations of total mercury in unfiltered and filtered water and ancillary data from Long-Term Monitoring Program Wells LTMP-3 and LTMP-5 in Operable Unit B-Terrestrial, Bremerton naval complex, Kitsap County, Washington, 2004–07.—Continued

[Data source: 2004–07 data, U.S. Navy (2007); 2008–09 data, U.S. Navy (2009); Dwight Leisle, U.S. Navy, written commun., 2010. **Final qualifier:** A, acceptable; J, estimated positive result; U, undetected; B, less than the reporting limit and greater than the instrument detection limit. **Abbreviations:** LTMP, Long-Term Monitoring Program; µg/L, microgram per liter; mg/L, milligram per liter; MP, measuring point; NGVD, National Geodetic Vertical Datum of 1929; MLLW, mean lower low water; –, not analyzed or does not apply]

Date collected	Time collected	Unfiltered mercury (µg/L)	Final qualifier	Filtered mercury (µg/L)	Final qualifier	Unfiltered aluminum (µg/L)	Final qualifier	Total suspended solids (mg/L)	Turbidity	Water level		
										Below MP	NGVD29	MLLW
						LTMP-5 (well OUBT-724)						
08-19-04	–	0.5	A	–	–	5	A	–	–	–	–	–
11-30-04	–	0.2	U	–	–	2	A	–	–	–	–	–
02-28-05	–	0.23	A	0.2	U	2	A	–	–	–	–	–
05-10-05	–	0.26	A	0.2	U	–	–	–	–	–	–	–
07-14-05	–	–	–	0.2	U	–	–	–	0	–	-0.78	5.21
10-18-05	–	5.24	A	–	–	5	U	0.96	0.96	–	3.42	9.41
01-19-06	–	0.2	U	–	–	–	–	0	0	–	-0.22	5.77
04-17-06	–	0.2	–	–	–	5	–	–	0	–	-2.5	3.49
10-10-06	–	0.2	BJ	–	–	30	–	21.3	21.3	9.58	2.34	8.33
10-11-07	–	0.2	U	–	–	–	–	43	19	10.68	1.24	7.23
10-15-08	[2]11:38	0.042	–	–	–	–	–	6	13	13.32	-1.4	4.59
04-01-09	[2]14:55	0.028	–	–	–	–	–	–	34	15.82	-3.9	2.09
N		11		3		5		5	8			
Average		0.67		0.2		9.4		14.3	11.0			
Standard deviation		1.52				11.6		18.2	12.9			
Minimum		0.028				2		0	0			
Maximum		5.24				30		43	34			

[1] One-third through ebbing cycle from higher low tide to higher high tide.

[2] At lower high tide.

Appendix B. Calculation of Sedimentation of Solids in Sinclair Inlet

The most common method used for estimating sedimentation rates involves carefully collecting a sediment core, measuring the mass of dry sediment for a series of vertical intervals in the core, and using low-radioactivity isotopes of lead (^{210}Pb or lead-210) as a time marker in the core. The sediment core first is collected in a manner that does not disturb the sediment-water interface. The core then is sectioned into numerous vertical intervals and multiple analyses are performed on each section. For most calculations of sedimentation rate, the mass of dry sediment for each vertical interval for a given area of sea floor, must be known. This measurement usually is made indirectly by measuring the percentages of water and dry solids in sediment from each vertical interval. The mass of sediment in each vertical section is added to the sum of the mass of solids in all sections above, to develop a curve of cumulative mass accumulation of solids with increasing depth in the core (fig. B1).

The timing component of a sedimentation rate is commonly determined by defining the decrease of very low radioactivity of ^{210}Pb (lead) with depth in the core due to radioactive decay. ^{210}Pb is produced in small amounts in the atmosphere from the decay of atmospheric ^{222}Rn (radon), and it is quickly scavenged onto atmospheric particles that settle onto water and land surfaces. Once in the water column of water bodies, ^{210}Pb tends to quickly settle to the sediment-water interface by attaching onto particles. Once delivered to the sediment column, the ^{210}Pb on the particles is buried by deposition of newly settling particles. ^{210}Pb decays with a half-life of 22.6 years. Assuming that the sedimentation rate is constant, the depth in the sediment at which the ^{210}Pb radioactivity decreases to half of the surface value represents 22.6 yr of sedimentation. The sedimentation rate based on ^{210}Pb activity is calculated from the slope of the linear regression of the natural log of the "unsupported activity" of ^{210}Pb against the cumulative mass accumulation of sediment:

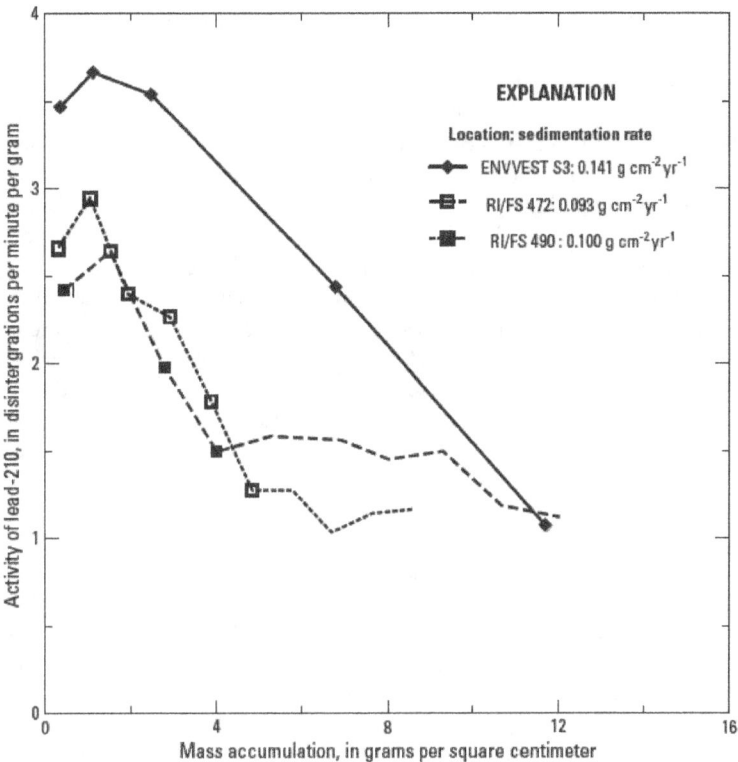

Figure B1. Lead-210 activity versus mass accumulation (locations 475 and 490) for cores from the Remedial Investigation/Feasibility Study and the nearby ENVironmental inVESTment project core (S3). The re-evaluated sedimentation rates for cores from locations 475 and 490 shown in the legend were calculated using only the part of the curve in which lead-210 activity decreases with increasing mass accumulation (data points with symbols).

$$\text{Sedimentation rate (gm cm}^{-2}\text{yr}^{-1}) \qquad (B1)$$
$$= -0.0311/\text{slope(log}(^{210}\text{Pb}) \text{ vs.}$$
$$\text{cumulative mass accumulation),}$$

where
    0.0311 is related to the radioactive half-life of
        ^{210}Pb (that is, decay rate).

The "unsupported activity" of ^{210}Pb is used because this measure corrects for the low levels of ^{210}Pb in a blank sample as well as the ^{210}Pb created by radium decay within the sediments. Note that the ^{210}Pb activity decreases down the core, so the slope of the logarithmic of unsupported ^{210}Pb is always a negative number.

Other indirect methods based of the vertical profile of a variety of constituents in the core are often used to corroborate the results of the ^{210}Pb method. The first appearance of elevated concentration of ^{137}Cs (cesium) in 1953 (as a result of atmospheric testing of thermonuclear weapons) is often used as a time marker in the sediments. Likewise, the peak concentration of ^{137}Cs in 1964, at the height of atmospheric testing before the Atmospheric Test Ban Treaty went into effect, is used as similar marker (Charles and Hites, 1987). Sometimes, the timing of increases of non-radioactive constituents to greater than natural levels can also be used as a time marker. In Sinclair Inlet, the increase of lead and mercury to greater than natural levels is associated with the opening of the Puget Sound Naval Shipyard (PSNS) in 1895. The better the agreement between various independent measurements, the more confidence one has in the value determined for the rate of sedimentation.

Three cores offshore of OU B Marine were collected in 1995 by the U.S. Navy as part of the RI/FS (U.S. Navy, 2002). Calculations based on the ^{210}Pb profiles of cores off Pier D (location 490, cited by Paulson and others, 2010) and off Pier 6 (location 475 cited by Paulson and others, 2010) reported sedimentation rates were about 0.42 g cm^{-2} yr^{-1}. However, the estimated rates were uncertain, because the ^{137}Cs concentration peak was much shallower than expected from the ^{210}Pb sedimentation calculations. One core (location 479 cited by Paulson and others, 2010) showed a dramatic break in the ^{210}Pb profile with depth.

Sedimentation rates in five cores along the center axis of Sinclair Inlet were measured in 2002 (Crecelius and others, 2003). The activity of ^{210}Pb in the sediment core sections was determined by a more sensitive method than that used in the RI/FS, in which the granddaughter ^{210}Pb alpha particle is counted on a silicon barrier diode detector. The sedimentation rates at the three cores offshore of OU B Marine (cores S2–S4 cited by Paulson and others, 2010) varied between 0.14 and 0.17 g cm^{-2} yr^{-1} (the top of core S4, collected adjacent to the area dredged in 2001 for the turning basin, was disturbed). In contrast, the sedimentation on the Gorst mudflats (core S1 cited by Paulson and others, 2010) was only 0.072 g cm^{-2} yr^{-1}, while the sedimentation rate on the eastern side of OU B Marine (S5) was 0.11 g cm^{-2} yr^{-1}. When the deepest vertical section where elevated total Pb and STHg concentrations were evident was assigned the date of 1895 (the opening of the PSNS), sedimentation rates for cores S1, S2, and S3 were comparable to the sedimentation rates calculated from the

activity of ^{210}Pb. However, this 1895 time marker method yielded higher sedimentations rates of 0.27 and 0.25 g cm^2 yr^1 for cores S4 and S5, respectively.

The sedimentation rates from the ENVVEST project were based on activity of ^{210}Pb and ranged between 0.072 and 0.17 g cm^{-2} yr^{-1}, whereas the rates from the Remedial Investigation/Feasibility Study were about 0.42 g cm^{-2} yr^{-1}. In order to investigate this discrepancy, the original ^{210}Pb activity of cores from locations 475 and 490 from the RI/FS were plotted against cumulative mass accumulation along with the ENVVEST project core (S3) nearest to the RI/FS cores (fig. B1). The two data sets show some similarities and differences. Both data sets show a mixed surface layer where the activity of ^{210}Pb, and thus the age of the sediments, remains fairly constant to a depth of about 4 cm (mass accumulation of about 2 g cm^{-2}). Both sets of data also show a similar rate of decrease in the activity of ^{210}Pb with increasing mass accumulation. However, the two data sets differ in the profile between 4 and 12 g cm^{-2}. The ENVVEST data show a consistent decrease in the activity of ^{210}Pb with increasing cumulative mass, while the Remedial Investigation data shows almost constant ^{210}Pb activity. This difference in the ^{210}Pb activity in the deeper part of the core is the reason why the calculated sedimentation rates from the two sets of cores differ.

The data set used to calculate the slope of ENVVEST cores included as many points below the mixed layer that provided a consistent decrease in the activity of ^{210}Pb with cumulative mass accumulation. In contrast, the data set used to calculate the slope of RI/FS cores contained data points from the entire core. Consequently, because the slope of the ^{210}Pb activity versus cumulative mass accumulation curve in the RI/FS cores flattens out above 4 g cm^{-2}, a larger sedimentation rate was estimated for the RI/FS cores. When the ^{210}Pb data from the RI/FS cores are reevaluated using data only from portion of the curve that decreases with increasing mass accumulation (RI/FS data points with symbols in fig. B1), the two data sets yield comparable sedimentation rates. Additionally, the smaller sedimentation rate calculated for ENVVEST core S3 is consistent with the time markers of ^{137}Cs data, whereas the larger sedimentation rates originally calculated by the RI/FS were not consistent with time markers of the ^{137}Cs data in the cores from locations 475 and 490. In addition, the mercury and lead data from the same ENVVEST cores and the assumption of the impact of the 1895 opening of the PSNS also are generally consistent with the lower estimated sedimentation rate.

In this study, the physical characteristics of sediment samples from the 500-ft and 1,500-ft LTMP grids were used to scale up the few sedimentation rates derived from ENVVEST project core data to the entire Sinclair Inlet study area. The following procedures were followed:

1. Zero sedimentation was assumed between mean lower low water (MLLW) and 5 m below MLLW.

2. The area below 5-m MLLW of the each 500-ft grid was calculated.

3. The area below 5-m MLLW of each 500-ft grid was assigned a sedimentation based on the average percentage of fines from the 2003, 2005 and 2007 LTMP.

   a. Grid cells with less than 50 percent fines were assigned a zero sedimentation rate.

   b. Grid cells with percentage fines between 50 percent and 85 percent were assigned a sedimentation rate of 0.09 g cm^{-2} yr^{-1} (average of S1 and S5 cores).

   c. Grid cells with percentage fines greater than 85 percent were assigned a sedimentation rate of 0.14 g cm^{-2} yr^{-1} (median of cores S2–S3).

4. Areas adjacent to the shoreline greater than 5-m MLLW, which were not covered by a 500-ft grid, were assigned a sedimentation rate based on the average of adjacent grid cells.

5. At the interface between the 500-ft grid and 1,500-ft grid, the area of 1,500-ft that was overlapped by a 500-ft grid was subtracted from the area of the 1,500-ft grid.

6. Areas at the interface between the 500-ft grid and 1,500-ft grid, which were not covered by either grid, were identified and assigned a sedimentation rate based on the percentage fines of adjacent 500-ft and 1,500-ft grids.

7. Areas adjacent to the shoreline shallower than 5-m MLLW, which were outside of the 500-ft grid and not covered by a 1,500-ft, were assigned a sedimentation rate based upon percentage fines of adjacent 1,500-ft grids.

8. All areas below 5-m MLLW and outside of the 500-ft grid are assigned a sedimentation rate based upon procedure. The area at the seaward boundary of the study areas, which were not covered by the 1,500-ft grid, were assigned a zero sedimentation rate because the median percentage of fines of the four grid cells is 42 percent fines.

9. The mass accumulation of all 500-ft grids, 1,500-ft grids, and areas outside of the both grids and adjacent to shorelines are summed to calculate a total mass sedimentation of Sinclair Inlet.

When the sedimentation rates assigned are applied to cells from the cells described above, an annual sedimentation of 7,120 metric tons of sediments (dry weight) is estimated over a depositional area of 6.23 km^2 (table B1). This compares to the ENVVEST estimate of 10,778 metric tons over an area of 7.9 km^2. Thus, it appears that the main difference between the approach described above and the ENVVEST approach, which average over different length scales, is that the depositional area in this study is smaller because of the exclusion of an area of coarse-grained sediments.

If shoreline areas within the BNC that were not covered by the 500-ft grid were added to the 500-ft BNC grid, annual sedimentation of 1,600 metric tons of sediment within the BNC is calculated. In contrast, 5,570 metric tons of sediments were annually deposited in greater Sinclair Inlet.

References Cited

Charles, M.J., and Hites, R.A., 1987, Sediments as archives of environmental pollution trends in Hites, R.A., and Eisenreich, S.J., eds., Sources and fate of aquatic pollutants: Washington, D.C., USA: American Chemical Society, p. 365–389.

Crecelius, E.A., Robert, K., Johnston, R.K., Leather, J., Guerrero, J., Mille, M., and Brandenberger, J., 2003, Contaminant mass balance for Sinclair and Dyes Inlets, Puget Sound, Washington, in Georgia Basin Puget Sound Research Conference, Vancouver, B.C., Canada, April 2, 2003, Proceedings: Olympia, Wash., Puget Sound Partnership, accessed April 2, 2012, at http://www.pugetsoundinstitute.org/archives/conf/2003GBPS_ResearchConference/SESSIONS/SESS_6E htm.

Paulson, A.J., Keys, M., and Scholting, K., 2010, Mercury in the Sediments, Waters and Biota of Sinclair Inlet, Puget Sound, Washington (1999–2007): U.S. Geological Survey Open-File Report 2009-1285, 220 p.

U.S. Navy, 2002, Final Remedial Investigation Report Operable Unit B, Bremerton Naval Complex, Bremerton, Washington: Final Report prepared by the URS Grier under contract #N62474-89-0-9295, Volume 1, 1,402 p.

Table B1.   Concentrations of carbon, and percent fines in sediment samples and assumed sedimentation rates from cells in Sinclair Inlet, Kitsap County, Washington, 2003–07.

[Values of mercury and assumed sedimentation in shore and interface cells are estimated from an average of adjacent cells. **Cell type:** GSI grid cells shown in figure 2; BNC grid cells shown in figure 3. Shore cells are unaccounted area along the shore that are between two grid cells. Interface cells are unaccounted area between BNC and GSI grid cells. **Fines**: Fraction of sediment less than 0.625 micrometer. **Abbreviations:** GSI, greater Sinclair Inlet; BNC, Bremerton naval complex; m^2, square meter; (g cm^{-2} yr^{-1}, gram per square centimeter per year]

Cell type	Cell No. or description	Total organic carbon (percentage by weight)		Fines (percentage by weight)		Area of cell (m^2)	Assumed sedimentation rate [g cm^{-2} yr^{-1}]	Annual sedimentation in cell (metric tons per year)
		Average	Standard deviation	Average	Standard deviation			
					Cells in greater Sinclair Inlet			
Grid	01	1.8	0.4	48	17	0	0	0
Grid	02	0.9	0.0	39	2	0	0	0
Grid	03	1.4	0.3	26	3	0	0	0
Grid	04	0.9	0.3	21	5	0	0	0
Grid	05	3.3	0.5	69	4	101,318	0.09	91
Grid	06	3.9	0.3	90	5	174,311	0.14	244
Grid	07	4.2	0.1	92	5	138,623	0.14	194
Grid	08	3.6	0.3	95	2	143,023	0.14	200
Grid	09	3.9	0.1	90	6	201,301	0.14	282
Grid	10	3.5	0.2	97	2	198,928	0.14	278
Grid	11	3.5	0.1	90	2	20,794	0.14	290
Grid	12	0.9	0.1	20	1	33,799	0	0
Grid	13	3.1	0.1	93	2	182,999	0.14	256
Grid	14	3.7	0.2	91	6	206,740	0.14	289
Grid	15	3.8	0.4	94	6	180,438	0.14	253
Grid	16	4.0	0.3	91	5	91,421	0.14	128
Grid	17	3.2	0.3	92	4	209,002	0.14	293
Grid	18	3.3	0.1	91	5	209,025	0.14	293
Grid	19	3.4	0.4	92	7	202,098	0.14	283
Grid	20	3.0	0.2	90	4	154,163	0.14	216
Grid	21	3.1	0.4	89	5	209,025	0.14	293
Grid	22	3.0	0.3	91	3	209,025	0.14	293
Grid	23	2.7	0.4	65	2	125,609	0.09	113
Grid	24	2.8	0.1	82	2	206,465	0.09	186
Grid	25	2.0	0.6	54	3	183,751	0.09	165
Grid	26	0.4	0.1	12	1	6,551	0	0
Grid	27	3.1	0.3	83	2	141,096	0.09	127
Grid	28	2.9	0.4	79	4	209,025	0.09	188
Grid	29	1.7	0.2	50	2	164,505	0.09	148
Grid	30	0.4	0.0	13	4	18,236	0	0
Grid	31	2.5	0.1	80	3	192,210	0.09	173
Grid	32	1.3	0.2	36	4	133,693	0	0
Shore	Between GSI cells 5 and 7					11,406	0.14	16
Shore	NW of GSI cells 5 and 6					22,967	0.09	21
Shore	Between GSI cells 19 and 23					12,523	0.09	11
Interface	NE of GSI cell 10 and SW of OU B Marine cells 1 and 2					19,330	0.14	27
Interface	NW of GSI cell 13 and between OU B Marine cells 2 and 5					22,378	0.14	31
Interface	Between GSI cells 13 and 17					15,917	0.14	22
Interface	NW of GSI cells 17 and 20, and SE of OU B Marine cells 13, 17 and 20					10,155	0.14	14
Interface	Between GSI cells 21 and 24, and SW of OU B Marine cells 37 and 38					78,255	0.14	110
Interface	NW of GSI cell 24 and between OU B Marine 28 and 44					16,920	0.09	15
Interface	Between GIS cells 27 and 31, and between OU B Marine cells 54 and 58					19,823	0.09	18
Interface	NW of GSI cell 31 and between OU B marine cells 58 and 62					6,135	0.09	6
GSI Total						4.67		5,570

Table B1.   Concentrations of carbon, and percent fines in sediment samples and assumed sedimentation rates from cells in Sinclair Inlet, Kitsap County, Washington, 2003–07.—Continued

[Values of mercury and assumed sedimentation in shore and interface cells are estimated from an average of adjacent cells. **Cell type:** GSI grid cells shown in figure 2; BNC grid cells shown in figure 3. Shore cells are unaccounted area along the shore that are between two grid cells. Interface cells are unaccounted area between BNC and GSI grid cells. **Fines:** Fraction of sediment less than 0.625 micrometer. **Abbreviations:** GSI, greater Sinclair Inlet; BNC, Bremerton naval complex; m^2, square meter; (g cm^{-2} yr^{-1}), gram per square centimeter per year]

Cell type	Cell No. or description	Total organic carbon (percentage by weight)		Fines (percentage by weight)		Area of cell (m^2)	Assumed sedimentation rate [g cm^{-2} yr^{-1}]	Annual sedimentation in cell (metric tons per year)
		Average	Standard deviation	Average	Standard deviation			
				Cells within OU B Marine				
Grid	01	3.1	0.3	60	14	23,216	0.09	20.89
Grid	02	3.3	0.2	93	5	23,225	0.14	32.51
Grid	03	1.7	0.1	44	6	13,896	0	0
Grid	04	3.0	0.3	91	9	23,225	0.14	32.52
Grid	05	2.8	0.1	88	6	23,225	0.14	32.51
Grid	06	2.7	0.2	87	5	17,675	0.14	24.74
Grid	07	2.7	0.2	90	5	23,225	0.14	32.51
Grid	08	2.5	0.1	88	3	23,225	0.14	32.51
Grid	09	2.6	0.1	91	1	23,225	0.14	32.51
Grid	10	1.4	0.3	52	7	22,528	0.09	20.27
Grid	11	2.3	0.1	87	3	23,225	0.14	32.52
Grid	12	2.6	0.4	90	0	23,225	0.14	32.51
Grid	13	2.9	0.2	88	6	23,224	0.14	32.51
Grid	14	1.0	0.3	32	5	7,767	0	0
Grid	15	2.2	0.0	88	2	21,240	0.14	29.74
Grid	16	2.5	0.2	90	2	23,225	0.14	32.52
Grid	17	2.7	0.1	92	4	23,225	0.14	32.51
Grid	18	1.6	0.8	42	5	12,435	0	0
Grid	19	2.3	0.2	81	5	23,225	0.09	20.9
Grid	20	2.7	0.2	93	1	23,225	0.14	32.51
Grid	21	1.8	0.2	68	5	22,576	0.09	20.32
Grid	22	3.0	0.2	97	2	23,225	0.14	32.51
Grid	23	2.9	0.2	90	5	23,225	0.14	32.51
Grid	24	3.1	0.4	88	4	13,106	0.14	18.35
Grid	25	3.0	0.3	85	6	23,225	0.14	32.51
Grid	26	3.2	0.4	83	11	23,225	0.09	20.9
Grid	27	3.3	0.4	85	3	23,225	0.14	32.51
Grid	28	2.3	0.6	71	13	22,111	0.09	19.9
Grid	29	3.5	0.1	83	11	23,225	0.09	20.9
Grid	30	3.0	0.1	86	1	23,225	0.14	32.52
Grid	31	2.9	0.1	93	1	23,225	0.14	32.52
Grid	32	2.7	0.1	93	2	23,225	0.14	32.52
Grid	33	3.2	0.3	88	2	16,225	0.14	22.72
Grid	34	2.6	0.3	72	15	23,225	0.09	20.9
Grid	35	2.5	0.3	76	3	23,225	0.09	20.9
Grid	36	3.0	0.1	88	3	23,225	0.14	32.52
Grid	37	3.0	0.3	87	2	23,124	0.14	32.37
Grid	38	2.2	0.1	69	3	23,225	0.09	20.9
Grid	39	2.1	0.8	37	6	10,331	0	0
Grid	40	2.7	0.3	73	5	22,921	0.09	20.63
Grid	41	3.1	0.6	74	10	20,728	0.09	18.66
Grid	42	2.7	0.1	79	3	21,315	0.09	19.18
Grid	43	2.9	0.5	72	6	15,329	0.09	13.7
Grid	44	2.9	0.1	90	4	23,225	0.14	32.51
Grid	45	3.0	0.7	66	22	18,044	0.09	16.24

Table B1. Concentrations of carbon, and percent fines in sediment samples and assumed sedimentation rates from cells in Sinclair Inlet, Kitsap County, Washington, 2003–07.—Continued

[Values of mercury and assumed sedimentation in shore and interface cells are estimated from an average of adjacent cells. **Cell type:** GSI grid cells shown in figure 2; BNC grid cells shown in figure 3. Shore cells are unaccounted area along the shore that are between two grid cells. Interface cells are unaccounted area between BNC and GSI grid cells. **Fines:** Fraction of sediment less than 0.625 micrometer. **Abbreviations:** GSI, greater Sinclair Inlet; BNC, Bremerton naval complex; m², square meter; (g cm⁻² yr⁻¹, gram per square centimeter per year]

Cell type	Cell No. or description	Total organic carbon (percentage by weight)		Fines (percentage by weight)		Area of cell (m²)	Assumed sedimentation rate [g cm⁻² yr⁻¹]	Annual sedimentation in cell (metric tons per year)
		Average	Standard deviation	Average	Standard deviation			
Cells within OU B Marine—Continued								
Grid	46	1.5	0.5	30	3	22,694	0	0
Grid	47	2.4	0.2	80	3	23,225	0.09	20.9
Grid	48	3.1	0.1	92	3	23,225	0.14	32.51
Grid	49	1.9	0.6	58	20	17,837	0.09	16.05
Grid	50	2.7	0.1	85	3	23,225	0.09	20.9
Grid	51	3.1	0.2	89	9	23,225	0.14	32.51
Grid	52	2.8	0.4	83	9	22,043	0.09	19.84
Grid	53	2.6	0.1	83	3	23,225	0.09	20.9
Grid	54	3.4	0.4	82	2	23,225	0.09	20.9
Grid	55	3.3	1.2	62	13	21,425	0.09	19.28
Grid	56	3.4	0.7	78	6	23,225	0.09	20.9
Grid	57	3.9	1.1	80	10	23,225	0.09	20.9
Grid	58	3.0	0.1	85	2	23,225	0.09	20.9
Grid	59	3.3	0.6	77	2	23,225	0.09	20.9
Grid	60	2.7	0.4	51	3	23,225	0.09	20.9
Grid	61	1.9	0.3	36	4	23,225	0.09	20.9
Grid	62	3.0	0.1	83	3	23,225	0.09	20.9
Grid	63	3.6	1.3	62	8	14,636	0.09	13.17
Grid	64	3.5	0.1	81	4	23,225	0.09	20.9
Grid	65	2.6	0.2	67	12	22,470	0.09	20.22
Grid	66	3.3	0.3	81	7	22,172	0.09	19.95
Grid	67	3.3	1.2	52	9	13,450	0.09	12.1
Grid	68	2.9	0.1	72	12	12,290	0.09	11.06
Grid	69	2.2	0.4	36	3	15,741	0	0
Grid	70	2.7	0.1	78	7	23,225	0.09	20.9
Grid	71	0.8	0.1	18	2	21,153	0	0
Interface	SW of OU B Marine cells 31 and 32 and NE of GSI cells 20 and 21					2,992	0.14	4.19
Interface	SW of OU B Marine 58 and W of GSI cell 31					4,898	0.09	4.41
Shore	Between OU B Marine cells 6 and 10					5,212	0.09	4.69
Shore	Between OU B Marine cells 24 and 28					5,681	0.09	5.11
Shore	Between OU B Marine cells 28 and 33					1,994	0.09	1.79
Shore	NE of OU B Marine cells 40 and 41					8,768	0.09	7.89
Shore	Between OU B Marine cells 37 and 42					6,835	0.09	6.15
Shore	NW of OU B Marine cells 59 and 63					11,944	0.09	10.75
BNC total							1.56	1,620
Sinclair Inlet total							6.23	7,190

Appendix C. Calculation of Total Mercury Concentrations of Solids

USGS data are compared (table C1) to U.S. Navy groundwater data and ENVVEST data for creeks, stormwater drains, and marine water using equation 1 to evaluate the representativeness of USGS samples.

The THg concentration of solids (ng/mg = mg/kg) collected from a terrestrial site (tables C2 and C3) or marine station (table C4) is obtained by dividing the concentration of PTHg (ng/L) obtained by filtering particles onto a QFF by the concentration of TSS (mg/L) obtained by filtering particles onto a polycarbonate filter (Huffman and others, 2012) according to equation 2. From December 2007 through September 2008, the water filtered through these two filters was taken from different bottles collected during the sequential filling of bottles at a site. If the TSS concentration in the water being sampled changed substantially during the sequential filling of bottles before October 2008, the calculation of THg of solids are biased. Beginning in October 2008, paired sets of solids for PTHg and TSS measurements were taken from the same bottle sampled at a site. Because of the possibility of sampling bias by sequential sampling during changing TSS conditions in 2008, PTHg concentrations associated with TSS concentrations less than 0.5 mg/L taken from a different bottle than the PTHg sample are disregarded in the statistical analyses.

Table C1. Comparisons of the sum of filtered and particulate total mercury concentrations with total mercury concentrations from whole water for comparable samples.

[**Abbreviations:** PTHg, particulate total mercury; FTHg, filtered total mercury; WTHg, whole water total mercury; ng/L, nanograms per liter; –, not available; <, less than]

Field identifer	PTHg + FTHg (ng/L)		WTHg (ng/L)		FTHg (ng/L)	
	Groundwater				USGS	LTMP
	USGS	LTMP	USGS	LTMP		
OUBT-722, pre-2008	–		980–6,690		–	210–970
OUBT-722, 2008-2009	500–2,124		1,190–3,690		453–1,970	–
OUBT-724, pre-2008	–		<200–5,240		–	<200
OUBT-724, 2008-2009	<10, 7.5		27.5–41.6		<10, 6.52	–

Creeks during dry baseflow		
	USGS	ENVVEST
Gorst Creek	0.53–1.16	2.89–4.01
Anderson Creek	0.75–1.50	1.2
Blackjack Creek	0.93–0.98	1.06–1.39
Annapolis Creek	0.78–1.20	1.13
Olney Creek	1.19–1.89	1.72

Stormwater drains during storms		
	USGS	ENVVEST
PSNS015	366	15–1,131
PSNS124	19.1	17.2–44.0
NAVY CITY	38.95	14.1–56.1

Upper layer marine water—Greater Sinclair Inlet		
	USGS	ENVVEST
Number of samples	29	17
Mean	0.89	1.09
Median	0.75	0.97
Minimum	0.45	0.72
Maximum	2.52	8.44

Upper layer marine water—OU B Marine		
	USGS	ENVVEST
Number of samples	23	10
Mean	0.84	2.36
Median	0.31	1.52
Minimum	0.02	0.78
Maximum	[1]1.91	10.71

[1]PTHg not measured. Assumed a value of 0.5 ng/L, medium value for layer.

Table C2.   Calculations of total mercury of solids discharging to greater Sinclair Inlet, Kitsap County, Washington, December 2007–January 2009.

[Values in **bold** are average of field replicate samples. **Abbreviations:** ng/L, nanogram per liter; mg/L, milligram per liter; THg, total mercury; ng/mg, nanogram per milligram and is same unit as mg/kg, milligram per kilogram; WWTP, wastewater treatment plant]

Field identifier	Date	Time	Particulate total mercury (ng/L)	Total suspended solids (mg/L)	THg of suspended solids (mg/kg)
Wastewater treatment plant					
Bremerton WWTP	05-08-08	0625	0.757	3.6	0.210
	07-15-08	0650	0.474	2.59	0.183
	08-14-08	0630	1.68	8.38	0.200
Stormwater					
Storm outfall at Sheridan Road near Bremerton	01-08-09	1720	1.83	11.65	0.157
Stormwater outfall at Port Orchard boat ramp	01-08-09	1840	2.85	20.63	0.138
Drainage outfall at Navy City near Gorst	01-07-09	2120	34.7	362.8	0.096
Streams					
Gorst Creek	05-09-08	1130	0.135	1.52	0.089
	07-14-08	0900	0.347	58.8	[1]0.006
Anderson Creel	05-08-08	1140	0.176	2.75	0.064
	07-15-08	1200	0.926	3.85	0.240
Blackjack Creek	05-07-08	1440	0.258	2.38	0.108
	07-14-08	1100	0.413	3.24	0.127
Annapolis Creek	05-09-08	1330	0.481	1.91	0.251
	07-15-08	0950	0.268	0.95	0.283
Olney Creek	05-08-08	1400	0.763	6.50	0.117
	07-14-08	1310	1.46	9.14	0.160

[1] Considered outlier.

Table C3. Calculations of total mercury of solids discharging to OU B Marine, Bremerton naval complex, Kitsap County, Washington, December 2007–March 2010.

[Values in **bold** are average of field replicate samples. **Abbreviations:** THg, total mercury; OUBT, Operable Unit B Terrestrial; OU NSC; Operable Unit Naval Supply Center; PSNS, Puget Sound Naval Shipyard; ng/L, nanogram per liter; mg/L, milligram per liter; ng/mg, nanogram per milligram and is same unit as mg/kg, milligram per kilogram]

Field identifier	Date	Time	Particulate total mercury (ng/L)	Total suspended solids (mg/L)	THg of suspended soids (mg/kg)
Dry dock drainage relief systems					
Dry Dock 1–5, pump 4	12-12-07	1115	6.53	0.49	[2]13.2
	12-12-07	[1]1245	2.79	0.74	3.75
	02-19-08	1030	2.60	0.74	3.52
Dry Dock 1–5, pump 5	01-29-08	1050	3.80	0.84	4.54
	03-26-08	1100	3.06	1.31	2.34
	04-23-08	1230	2.07	0.64	3.23
	05-28-08	1050	2.12	1.47	1.45
	06-24-08	1030	57.4	43.62	1.32
Dry Dock 6	02-19-08	[1]1600	4.38	0.85	5.17
	02-19-08	1630	2.03	0.58	3.49
	03-26-08	1300	2.57	0.29	[2]8.74
	04-25-08	1340	2.90	0.16	[2]17.7
	05-28-08	1200	1.94	1.43	1.36
	06-24-08	1120	1.75	1.37	1.27
Steam plant effluent					
Steam Plant	03-26-08	1400	59.4	1.45	40.9
	04-25-08	1030	32.5	0.47	68.72
	05-28-08	1350	23.85	1.39	17.1
	06-24-08	1220	2.16	0.73	2.95
Groundwater in Operating Units B and NSC captured by the dry dock drainage relief systems					
OUBT-406R	01-31-08	1500	0.549	0.58	0.94
OUBT-406R	04-23-08	1500	<0.057	0.56	<0.101
OUBT-709	01-30-08	1550	8.39	0.18	46.82
OUBT-709	04-24-08	1530	3.57	1.13	3.15
OUBT-724	02-01-08	1030	1.79	1.00	1.78
OUBT-724	04-24-08	1200	0.925	2.20	0.42
OUNSC-380	01-30-08	1310	0.583	0.43	1.35
OUNSC-380	04-22-08	1030	0.667	0.55	1.22
Stormwater					
PSNS124.1	01-07-09	1720	32.7	158	0.21
PSNS124	01-07-09	1830	17.6	35.53	0.495
PSNS015	01-07-09	2010	222	151.8	1.46
Tidal flushing of PSNS 015 storm drain					
PSNS015-2253	12-29-09	1340–2000	21.89	1.35	16.2
PSNS015-2253	03-31-10	1010–1400	54.57	2.86	19.1

[1] Composite sample, previous 24 hours.

[2] Total suspended solids concentrations less than 0.5 mg/L and taken from a different bottle than the PTHg sample. May be affected by temporal alaising and disregarded in statistical applications and plots.

Table C4.   Calculations of total mercury of solids collected in the marine waters of Sinclair Inlet, Kitsap County, Washington, August 2008–August 2009.

[Values in **bold** are the average from two different bottles. **Abbreviations:** ng/L, nanogram per liter; mg/L, milligram per liter; THg, total mercury; ng/mg, nanogram per milligram and is same unit as mg/kg, milligram per kilogram; –, not analyzed]

Field identifier	Date	Time	Particulate total mercury (ng/L)	Total suspended solids (mg/L)	THg of suspended solids (mg/kg)
Bremerton naval complex lower layer					
BNC-39-BOT	08-18-08	1110	4.13	4.03	1.025
BNC-39-BOT	02-06-09	1100	0.745	1.22	0.611
BNC-39-BOT	06-05-09	1000	1.7	2.8	0.607
BNC-39-BOT	08-07-09	0900	5.59	**6.58**	0.849
BNC-52-BOT	08-15-08	0930	0.392	1.13	0.347
BNC-60-BOT	02-04-09	1300	0.634	1.73	0.366
BNC-60-BOT	06-03-09	0930	1.21	2.82	0.429
BNC-60-BOT	08-05-09	0900	0.641	1.07	0.599
BNC-71-BOT	08-18-08	1240	0.767	101.2	0.008
BNC-71-BOT	02-06-09	1300	0.281	6.25	0.045
BNC-71-BOT	06-05-09	1200	0.717	2.27	0.316
BNC-71-BOT	08-07-09	1300	1.02	**1.97**	0.517
Bremerton naval complex upper layer					
BNC-39-SURF	02-06-09	1130	0.401	0.71	0.565
BNC-39-SURF	06-05-09	1000	0.422	2.26	0.187
BNC-39-SURF	06-05-09	1001	0.452	2.72	0.166
BNC-39-SURF	08-07-09	0900	0.672	3.42	0.196
BNC-39-SURF	08-18-08	1020	0.629	2.01	0.313
BNC-52-SURF	08-15-08	1110	0.264	2.16	0.122
BNC-52-SURF	09-19-08	1240	0.566	4.35	0.130
BNC-52-SURF	10-22-08	1130	0.456	1.09	0.418
BNC-52-SURF	11-18-08	1200	0.528	1.21	0.436
BNC-52-SURF	02-04-09	1130	0.302	2.59	0.117
BNC-52-SURF	03-18-09	1230	0.283	0.92	0.308
BNC-52-SURF	04-08-09	1100	0.435	**2.99**	0.145
BNC-52-SURF	05-07-09	1100	0.522	**1.69**	0.308
BNC-52-SURF	06-03-09	1100	0.165	2.02	0.082
BNC-52-SURF	07-08-09	1130	0.175	**2.45**	0.071
BNC-52-SURF (set 1)	08-05-09	1030	0.517	**4.24**	0.122
BNC-52-SURF (set 1)	08-05-09	1032	0.667	**6.45**	0.103
BNC-71-SURF	08-18-08	0940	0.38	0.97	0.392
BNC-71-SURF	02-06-09	1330	0.216	0.73	0.296
BNC-71-SURF	06-05-09	1230	0.573	3.15	0.182
BNC-71-SURF	08-07-09	1300	0.513	3.22	0.159
Convergence zone					
CZ-BOT	02-05-09	1230	0.462	1.84	0.251
CZ-SURF	08-14-08	1240	0.303	2.56	0.118
CZ-SURF	09-19-08	1140	0.443	1.9	0.233
CZ-SURF	10-22-08	1050	0.502	1.06	0.474
CZ-SURF	11-18-08	1110	0.586	1.2	0.488
CZ-SURF	02-02-09	1130	0.339	0.76	0.446
CZ-SURF	03-17-09	1100	0.427	1.45	0.294
CZ-SURF	04-07-09	1100	0.742	2.44	0.304
CZ-SURF	05-06-09	1100	0.587	**2.15**	0.273
CZ-SURF	06-01-09	1100	0.527	2.88	0.183
CZ-SURF	07-07-09	1130	0.394	**1.95**	0.203
CZ-SURF	08-03-09	1130	0.482	**2.34**	0.206

Table C4. Calculations of total mercury of solids collected in the marine waters of Sinclair Inlet, Kitsap County, Washington, August 2008–August 2009.—Continued

[Values in **bold** are the average from two different bottles. **Abbreviations:** ng/L, nanogram per liter; mg/L, milligram per liter; THg, total mercury; ng/mg, nanogram per milligram and is same unit as mg/kg, milligram per kilogram; –, not analyzed]

Field identifier	Date	Time	Particulate total mercury (ng/L)	Total suspended solids (mg/L)	THg of suspended solids (mg/kg)
Greater Sinclair Inlet lower layer					
SI-IN-BOT	08-15-08	1400	0.633	1.26	0.502
SI-IN-BOT	02-04-09	1430	0.416	0.96	0.433
SI-IN-BOT	06-03-09	1330	0.791	2.2	0.360
SI-IN-BOT	08-05-09	1230	1.4	**3.42**	0.409
SI-OUT-BOT	08-14-08	1000	1.79	3.71	0.482
SI-OUT-BOT	02-02-09	1400	0.821	1.95	0.421
SI-OUT-BOT	06-01-09	1400	0.634	2.58	0.246
SI-OUT-BOT	08-03-09	1230	0.184	**2.12**	0.087
SI-PO-BOT	08-14-08	1400	0.291	2.18	0.133
SI-PO-BOT	02-02-09	0950	0.845	1.93	0.438
SI-PO-BOT	06-01-09	0900	1.94	3.42	0.567
SI-PO-BOT	08-03-09	0930	0.478	**1.29**	0.372
Greater Sinclair Inlet upper layer					
SI-IN-SURF	08-15-08	1200	0.475	3.98	0.119
SI-IN-SURF	09-18-08	1120	0.785	10.25	0.077
SI-IN-SURF	10-21-08	1100	0.718	**7.22**	0.100
SI-IN-SURF	04-08-09	1150	0.434	**2.14**	0.203
SI-IN-SURF	05-07-09	1200	0.826	2.19	0.376
SI-IN-SURF	06-03-09	1130	0.153	2.13	0.072
SI-IN-SURF	07-08-09	1100	0.196	**1.85**	0.106
SI-IN-SURF	08-05-09	1200	1.79	**10.68**	0.168
SI-IN-SURF	11-17-08	1120	0.477	1.3	0.367
SI-IN-SURF	02-04-09	1040	0.399	0.98	0.407
SI-IN-SURF	03-18-09	1130	0.34	1.44	0.236
SI-OUT-SURF	02-02-09	1420	0.328	0.78	0.421
SI-OUT-SURF	06-01-09	1330	0.239	2.62	0.091
SI-OUT-SURF	08-03-09	1230	0.572	**4.11**	0.139
SI-OUT-SURF	08-14-08	1140	0.346	3.78	0.092
SI-OUT-SURF	08-14-08	1150	0.25	–	0.066
SI-PO-SURF	09-18-08	1240	0.489	**4.99**	0.098
SI-PO-SURF	09-18-08	1241	0.625	–	0.125
SI-PO-SURF	10-21-08	1205	0.438	0.93	0.471
SI-PO-SURF	11-17-08	1150	0.574	1	0.574
SI-PO-SURF	02-02-09	1050	0.317	0.44	0.720
SI-PO-SURF	03-17-09	1200	0.403	1.37	0.294
SI-PO-SURF	04-07-09	1200	0.453	2.61	0.174
SI-PO-SURF	05-06-09	1200	0.309	**1.45**	0.214
SI-PO-SURF	06-01-09	1200	0.372	2.7	0.138
S-PO-SURF	07-07-09	1200	0.357	**2.74**	0.130
S-PO-SURF	08-03-09	1040	0.33	2.62	0.126

www.ingramcontent.com/pod-product-compliance
Lightning Source LLC
Chambersburg PA
CBHW081507170526
45166CB00008B/2572